EDIBLE ID

FOOD AS CULTU

M000290893

Heritage, Culture and Identity

Series Editor: Brian Graham,
School of Environmental Sciences, University of Ulster, UK

Edible Identities:
Food as Cultural Heritage

Edited by

RONDA L. BRULOTTE
University of New Mexico, USA

MICHAEL A. DI GIOVINE
West Chester University, USA

LONDON AND NEW YORK

First published 2014 by Ashgate Publishing

Published 2016 by Routledge
2 Park Square, Milton Park, Abingdon, Oxfordshire OX14 4RN
711 Third Avenue, New York, NY 10017, USA

First issued in paperback 2016

Routledge is an imprint of the Taylor & Francis Group, an informa business

British Library Cataloguing in Publication Data
A catalogue record for this book is available from the British Library.

The Library of Congress has cataloged the printed edition as follows:
Brulotte, Ronda L.
  Edible identities : food as cultural heritage / By Ronda L. Brulotte and Michael A. Di Giovine.
    pages cm. -- (Heritage, culture and identity)
  Includes bibliographical references and index.
  ISBN 978-1-4094-4263-9 (hardback)
  1. Food habits. I. Di Giovine, Michael A. II. Title.
  GT2850.B78 2014
  394.1'2--dc23

                                                                2014012565

ISBN: 978-1-138-63494-7 (pbk)
ISBN: 978-1-4094-4263-9 (hbk)

Contents

List of Figures

For our parents and grandparents, who instilled in us an appreciation of food and heritage.

Notes on Contributors

Warren Belasco is Professor Emeritus of American studies at the University of Maryland, Baltimore County, as well as Visiting Professor of Gastronomy at Boston University. He is the author of *Meals to Come: A History of the Future of Food* (University of California Press, 2006), *Appetite for Change: How the Counterculture Took on the Food Industry* (Pantheon Books, 1989, Cornell University Press, 1993, 2007), and *Food: The Key Concepts* (Berg, 2008). He co-edited *Food Nations: Selling Taste in Consumer Societies* (Routledge, 2002), *The Oxford Encyclopedia of Food and Drink in America* (Oxford University Press, 2004), and *Food Chains: From Farmyard to Shopping Cart* (University of Pennsylvania Press, 2009).

Regina F. Bendix is Professor of Cultural Anthropology/European Ethnology at the University of Göttingen, Germany. A native of Switzerland, she is co-editor of the journal *Ethnologia Europaea* and the author of several books and many articles about European and American folklore and folklife, including *Backstage Domains: Playing William Tell in Two Communities* (1989), *In Search of Authenticity* (1997) and, co-edited with Galit Hasan Rokem, *A Companion to Folklore* (2012).

Sara M. Bergstresser is a medical and cultural anthropologist, and she is currently a Visiting Lecturer in Anthropology at Boston University. Her research addresses the intersection of health and society, including mental health policy and stigma, global bioethics, religion and health, disability studies, and science and technology studies. She holds a Ph.D. in Anthropology from Brown University and an M.P.H. from Harvard University.

Ronda L. Brulotte is Assistant Professor of Anthropology and an affiliated faculty member of the Latin American and Iberian Institute at the University of New Mexico. She earned a Ph.D. in Anthropology (2006) and M.A. in Latin American Studies (1999) at the University of Texas at Austin. Her teaching and research areas include tourism, material culture, art and aesthetics, and the anthropology of food. Her book *Between Art and Artifact: Archaeological Replicas and Cultural Production in Oaxaca, Mexico* (University of Texas Press, 2012) is an ethnographic examination of the politics of heritage tourism and artisan production in southern Mexico. She is currently working on a project on agave cultivation and mezcal production in Oaxaca, Mexico.

Erick Castellanos is Associate Professor of International Studies and Anthropology at Ramapo College of New Jersey. His research interests lie in the intersection of food, culture, and identity in a transnational context. He has conducted fieldwork in Italy, among Latinos in the U.S., and in Mexico. He received his Ph.D. in Anthropology from Brown University in 2004 and has a M.A. in International Relations from Johns Hopkins-SAIS.

Carole Counihan is Professor Emerita of Anthropology at Millersville University. She is author of *Around the Tuscan Table: Food, Family and Gender in Twentieth Century Florence* (Routledge, 2004) and *The Anthropology of Food and Body: Gender, Meaning and Power* (Routledge, 1999). She is editor of *Food in the USA: A Reader* (Routledge, 2002); with Penny Van Esterik of *Food and Culture: A Reader* (Routledge, 1997, 2008, 2013); with Valeria Siniscalchi of *Food Activism: Agency, Democracy and Economy* (Bloomsbury, 2014); and with Psyche Williams-Forson of *Taking Food Public: Redefining Foodways in a Changing World* (Routledge, 2012). She received a 2005–2006 National Endowment for the Humanities Fellowship to complete her book *A Tortilla Is Like Life: Food and Culture in the San Luis Valley of Colorado* (University of Texas Press, 2009). She is also the editor-in-chief of the scholarly, interdisciplinary, international journal *Food and Foodways*.

Greg de St. Maurice is a Ph.D. candidate in Cultural Anthropology at the University of Pittsburgh. He has a M.Sc. in Social Anthropology from Oxford University and M.A. degrees in International Relations from American University and Ritsumeikan University. His primary research interests include place-making, food and agriculture, and globalization.

Michael A. Di Giovine is Assistant Professor of Anthropology at West Chester University of Pennsylvania and Honorary Fellow in the Department of Anthropology at the University of Wisconsin-Madison. A former tour operator, his research in Europe and Southeast Asia focuses on global mobilities (tourism/ pilgrimage and immigration), heritage, foodways, and religion. He earned a B.S. in Foreign Service from Georgetown University, and both an A.M. in the interdisciplinary social sciences and a Ph.D. in socio-cultural anthropology from the University of Chicago; his 2012 dissertation explored the cult of St. Padre Pio of Pietrelcina. A founding board member of both the American Anthropological Association's Anthropology of Tourism Interest Group and the Tourism-Contact-Culture research network, Michael sits on the American Anthropological Association's Task Force on Cultural Heritage, the academic board of *The Journal of Tourism and Cultural Change* and *The International Journal of Religious Tourism and Pilgrimage*, and is the book reviews editor for *Journeys: The International Journal of Travel and Travel Writing*. Michael is the author of *The Heritage-scape: UNESCO, World Heritage, and Tourism* (Lexington Books, 2009), and co-editor (with David Picard) of *Tourism and* *the Power of Otherness: Seductions of Difference* (Channel View, 2014). He is the series co-editor of *Anthropology of Tourism: Heritage, Mobility and Society* (Lexington Books).

Josep Maria Garcia-Fuentes is Lecturer in Architecture at the School of Architecture, Planning and Landscape at Newcastle University, UK. He is also Adjunct Professor of Architecture at the *Universitat Politècnica de Catalunya-BarcelonaTECH*, where he has served as Vice Dean and International Coordinator at the *Escola Tècnica Superior d'Arquitectura del Vallès-Barcelona*, Spain (2011–2014). He was awarded the First National Prize of Spain for university graduates (2006), and his work has been supported by the *Caja de Arquitectos* (2004), the Government of Spain (2007–2011), and the Samuel H. Kress Foundation for the Society of Architectural Historians (2011) and the Santander Bank (2014). His research issues are about the history of architecture, urban history, heritage making processes, and identity, as well as preservation, focusing on the nineteenth and twentieth centuries. He has published and spoken widely about them at international conferences and is currently working on the publication of his Ph.D. dissertation entitled *The Construction of the Modern Montserrat*. He is also member of the funded research project on the Barcelona's System of Market Halls led by Manel Guàrdia.

Cristina Grasseni, M.Phil., Ph.D., is Associate Professor of Cultural Anthropology at Utrecht University. Her research interests include alpine heritage, local development, the anthropology of food, and provisioning activism. Her publications include: *Beyond Alternative Food Networks. Italy's Solidarity Purchase Groups* (New York, Bloomsbury, 2013), *Developing Skill, Developing Vision. Practices of Locality at the Foot of the Alps* (Oxford, Berghahn, 2009) and *Skilled Visions. Between Apprenticeship and Standards* (as editor, Oxford, Berghahn, 2007).

Manel Guàrdia is Professor of Architecture and Urban History at the *Universitat Politècnica de Catalunya–BarcelonaTECH*. He has been Director of the Department of Architectural Composition (2007–2012) and Director of the Ph.D. Program on History and Theory of Architecture at the same university (2007–2012). His publications and research issues, presented at international academic conferences and published in academic journals, include architecture, urban history, the history of urbanism and Barcelona's urban history. He is currently leading a research project on Barcelona's System of Market Halls funded by the Government of Spain. He is the co-author (with José Luis Oyón) of the volume (in Spanish) entitled *Making Cities through Market Halls: Europe, 19th and 20th Centuries* (Museu d'Història de la Ciutat, Ajuntament de Barcelona, 2010). He was also the co-editor (with Javier Monclús) of the volume *Culture, Urbanism and Planning* (Ashgate, 2006).

Miha Kozorog is a researcher at the Department of Ethnology and Cultural Anthropology at the University of Ljubljana. His main research topics include popular culture, tourism, and place-making. He has conducted ethnographic fieldwork in Slovenia and Italy. He published two monographs (in Slovenian): *Anthropology of an Emergent Tourist Destination: Place, Festivals and Local Identity in Tolminska Region, Slovenia* (2009) and *Festival Places: Concepts, Politics and Hope at the Periphery* (2013).

José Luis Oyón is Professor of Planning and Urban History at the *Universitat Politècnica de Catalunya–BarcelonaTECH*. His publications and research issues published in academic journals include urban history, specially Barcelona's urban and social history. He is the co-author with Manel Guàrdia of the volume *Making Cities through Market Halls: Europe, 19th and 20th Centuries* (Museu d'Història de la Ciutat, Ajuntament de Barcelona, 2010). He was also the co-editor (with Javier Monclús and Manel Guàrdia) of the volumes of France and Iberian Peninsula of *Atlas histórico de ciudades europeas* (Salvat-Hachette, 1994, 1996). His research on Barcelona working-class urban history has been published as *La quiebra de la ciudad popular: espacio urbano, inmigración y anarquismo en la Barcelona de entreguerras, 1914–1936* (El Serbal, Barcelona 2009).

Heather Paxson is Associate Professor of Anthropology at the Massachusetts Institute of Technology, where she teaches courses on food and on craft practice. *The Life of Cheese: Crafting Food and Value in America*, her ethnography of artisan cheesemaking in the United States, was published in 2013 with the University of California Press, which also published her first book, *Making Modern Mothers: Ethics and Family Planning in Urban Greece* (2004).

Clare A. Sammells is Assistant Professor of Anthropology at Bucknell University. She holds a doctorate from the University of Chicago and her research interests include economic anthropology, tourism, and the anthropology of food, with a focus on Bolivia. She is the co-editor (with Helen R. Haines) of *Adventures in Eating: Anthropological Tales of Dining around the World* (University of Colorado Press, 2010).

Alvin Starkman holds a J.D. from Osgoode Hall Law School in Toronto (1984) and an M.A. in anthropology from York University (1978). A permanent resident of Oaxaca, Mexico, since 2004, he has written over 300 articles about life and cultural traditions in southern Mexico as well as legal opinions, for literary, scholarly and a broad diversity of travel, crafts and culinary publications. His areas of specific interest in Oaxaca lie with its culinary heritage, fermented and distilled beverages, and sustainable living models. He consults to academics working in Oaxaca, domestic and foreign interests regarding mezcal, pulque and local cuisine, documentary film companies, and federal initiatives such as Mexico Today (*Marca País—Imagen de México*) and CONOCER (*Consejo Nacional* *de Normalización y Certificatión de Competencias Laborales*). He is the author of *Mezcal in the Global Spirits Market: Unrivalled Complexity, Innumerable Nuances* (Carteles Editores – P.G.O., 2014).

Susan Terrio is Professor of Anthropology and French Studies at Georgetown University. She is the author of *Crafting the Culture and History of French Chocolate* published by the University of California in 2000. She currently holds a joint appointment in the Department of Anthropology and the Department of French and served as the inaugural chair of the Department of Anthropology from 2008–2011. She is a 2012–2013 Fellow at the Woodrow Wilson International Center for Scholars in Washington where she is writing a book on unaccompanied minors in the United States. Her areas of expertise include the anthropology of contemporary France, Western Europe and the U.S. as well as the social and cultural history of France since the revolution of 1789. Specific interests center on social class and educational systems, craft and commoditization, food and foodways, race and ethnicity, migration, law, and juvenile and immigration courts in France and the U.S.

Psyche Williams-Forson is Associate Professor of American Studies at the University of Maryland College Park and an affiliate faculty member of the Women's Studies and African American Studies departments and the Consortium on Race, Gender, and Ethnicity. She is co-author with Carole Counihan of *Taking Food Public: Redefining Foodways in a Changing World* (Routledge, 2011) and *Building Houses Out of Chicken Legs: Black Women, Food, and Power* (University of North Carolina Press, 2006). Her new research explores the role of the value market as an immediate site of food acquisition and a project on class, consumption, and citizenship among African Americans by examining domestic interiors from the late nineteenth century to the early twentieth century.

Introduction

Food and Foodways as Cultural Heritage

Michael A. Di Giovine and Ronda L. Brulotte

Anthropologists and other scholars sensitive to the world's diversity of cultural forms and practices are often loath to speak of universalities. However, along with sex and death (and perhaps taxes, as the old adage goes), the production, elaboration, and consumption of food may very well be one of those sets of processes that are common to all human beings. Just as we humans must procreate, and just as our bodies will all eventually pass from this state of living, we all must eat to sustain ourselves. Yet how we eat, and what we eat, and when we eat, and with whom we eat, all uniquely vary from place to place, group to group, time to time—thanks to longstanding geographic, economic, social, and cosmological differences throughout the world. And within these discrete social entities, food binds people together; it is individually consumed, entering into our singular bodies, but often communally grown, processed, and prepared. To feed a village, it takes a village—or, in this age of globalization and industrial food, many villages—and, as such, food is often a primary marker of individual and group identity. Food is therefore extremely affective; its taste on our individual tongues often incites strong emotions, while the communal, commensal experience of such sensations binds people together, not only through space but time as well, as individuals collectively remember past experiences with certain meals and imagine their ancestors having similar experiences. When this occurs, food is transformed into heritage.

In the grand scheme of things, heritage is a relatively new term, its origins stemming from the French and British imperial eras to denote the accumulation of wealth or patrimony (*patrimoine*) of tangible and intangible goods that a society inherits from the past (*héritage*), preserves in the present, and passes on to the future. These are mediators, linking members of society together through space and time, serving as referential touchstones for a group's self-identification—often at the level of the nation (Hall 1999)—and representing the group to outsiders (Di Giovine 2009a). Yet while governments, preservationists, and cultural resource managers may refer to "their" heritage or "the" heritage as if it were discrete set of goods or practices that can be unproblematically claimed or possessed, scholars have argued that heritage is less an identifiable thing than a constructed discourse strategically deployed for political, economic, or ideological goals (see for example Di Giovine 2009a, Handler 1988, Kirshenblatt-Gimblett 1998). What may be one's heritage may not be another's, but also what may be claimed as one group's unique heritage might also be claimed, or contested, by other groups.

Already affective in nature, food therefore takes on even greater emotional weight when designated as "heritage." This heritage can be small in scale, demarcating a particular group or community; it can likewise be large in scale, attempting to solidify nationalistic ideologies or multicultural ideals that purport to unify, homogenize, or celebrate cultural diversity. In all of these cases, it is clear that heritage imparts particular value claims on people, their histories, social structures, and traditions. It is also able to contain and embody the memories of people and places, across space and time—becoming, in the words of Barbara Kirshenblatt-Gimblett, "edible chronotopes" (2004a: xiii). It is in this way that food as cultural heritage is "good to think," to borrow from Lévi-Strauss' oft-cited aphorism (Lévi-Strauss 1963: 89).

This volume examines the ways in which the cultivation, preparation, and consumption of food is used to create identity claims of "cultural heritage" on local, regional, national and international scales. In particular, the chapters deal with how food is used to mark insiders and outsiders within ethnic groups, how the same food's meanings change within a particular society based on class, gender or taste; and how traditions are "invented" for the economic and social revitalization of communities. Contributors collectively delve into the problematic of designating, classifying, and valorizing foods as a particular culture's heritage. This issue is extraordinarily timely, as an increasing number of regions (particularly in the European Union) are creating *terroir* designations for broad classes of food, and as UNESCO has begun to institutionally designate the food and cuisines of several European, Middle Eastern and Latin American countries as "intangible cultural heritage." Along with this is the growth in worldwide culinary tourism—travel that is explicitly centered on exploring, experiencing, and enjoying the cuisine of a foreign place—a way of engaging with alterity in a standardized way (Long 2004, cf. Picard and Di Giovine 2014), and often, as Clare Sammells (in this volume) and others have argued (e.g. Long 2004), for groups of people to play out their self-identifications as "cosmopolitan" or, as Johnston and Baumann intimate (2010), as cosmopolitan "foodies."

Certainly there is a strong commercial undercurrent to designating certain foods as cultural heritage, as many of the contributions in this volume make abundantly clear. Indeed, the so-called "heritage industry" (Hewison 1987) has been accused of commodifying, museumifying, and commercializing cultures and their material expressions since the inception of scholarly writing on the subject (e.g. Lowenthal 1998). Likewise, the global food industry is increasingly looked on skeptically by scholars as it homogenizes, stereotypes, and "MacDonaldizes" (Ritzer 2012) not only the food itself, but the experience of consuming particular foods, and the imaginaries of the alimentary culture to which it is related (cf. Watson 2006, Wilk 2006, Ram 2007, Schlosser 2012). It is also important to remember that foodways themselves are always embedded in wider economic systems, even in the most rural of settings: even the most self-sufficient farmer nearly always must pay for some component necessary for the cultivation of a crop—such as rent or taxes on the land tilled, the water or machinery used, the feed, fertilizer, or pesticides necessary to ensure a healthy yield, or the labor (even temporary) employed to cultivate and harvest the goods. Food is then often gifted, donated, turned into payment for laborers, exchanged for other foods, or sold in markets—in short, it is turned into commodities, wages, and objects. The chapters in this volume thus acknowledge the commercial advantages that heritage designations on food are perceived to provide; they also interrogate the connections between heritage and food industries—as well as the "culture" industries and especially the travel industry. Yet they also view the connection between food and heritage to be deeper, more complex, and multifaceted. That is, they reveal that, to see merely the commercial in both the heritage and food "industries," particularly on the local scale, is to neglect to see how they act as mediators, bringing people together through economic exchange (Mauss 2000), and how, as a foodstuff travels through a foodway, and an object is transformed into heritage, it is used to indicate, explicate and replicate important ideological claims on identity, ownership, sovereignty, and value.

Food and Ethnic Identity

One of the major themes of these contributions is how food, conceptualized as a marker of heritage, creates and reinforces ethnic group identity in increasingly multicultural milieux. The term "multicultural" has become a buzzword for post-colonial nation-states that seek to homogenize or otherwise mitigate tensions between diverse ethnic groups within their societies (Asad 1993). But we can also extend this notion more broadly to the world system (cf. Mintz 1986), particularly in this age of globalization in which spatial and temporal distances are conflated and actors entertain imaginaries of being part of a cohesive whole (see Giddens 1991, Robertson 1992)—a "culture of cultures" as anthropologist Marshall Sahlins called it (2000: 488), or an Aristotelian *oikumene* of sorts. This is especially the case in World Heritage discourses by UNESCO and its affiliates, who, through the meta-narrative of "unity in diversity" emphasize that the one commonality all cultures possess is difference (Di Giovine 2009a: 119–44); it also implicitly operates among nation-states linked in world systems of trade and commerce (Mintz 1986, Castellanos and Bergstresser this volume), through immigration, travel, diplomacy and even war (see Anderson 1993).

Like heritage, ethnicity is not a fixed thing but a process of creating and reproducing classificatory distinctions between in-groups and out-groups by people who perceive themselves as distinct from others (Eriksen 1992: 3). As anthropologist Fredrik Barth famously posited, these categories of belonging are created and maintained by well-defined boundaries that demarcate ethnic insiders and outsider Others (1969: 10). While, as Psyche William-Forson's contribution to this volume shows, these boundaries are in reality quite porous (see also Bashkow 2004: 444), they are generally understood as historical and naturalized, and are represented by both parties through shared stereotypes (Barth 1969: 19) and imaginaries that dialectically modify themselves over time. These stereotypes can include tangible elements—such as physiognomic or bodily features, or dress—as well as immaterial elements—such as language, cultural knowledge, shared worldviews and values, religion, shared history, symbolic practices, and other forms of ethnic performances labeled as "tradition." Importantly, however, these stereotypes are variable in semiotic importance even within a particular group, depending on the individual person or situation—such that Bakalian goes as far as describing ethnicity itself as a "state of mind," fueled by nostalgia (Bakalian 1993: 13, quoted Nagel 1994: 154). Indeed, in commenting on Barth's formalist approach, Eller also emphasizes a mental component when he argued that ethnicity is predicated on a "consciousness of difference" (Eller 1999: 9) that is highly dependent on the cultural contexts in which a particular segment of the ethnic group finds itself. Therefore, the construction, persistence, and replication of ethnic identity occur not through a linear process of presentation and reception, but rather through a relational dialectic of negotiation, contestation, self-fashioning, and re-presentation (see Lu and Fine 1995: 535; cf. Alba 1990: 75; Chow 1993, Denzin 1994).

Early in the development of the burgeoning discipline of food studies, social scientists have pointed to the importance of food in facilitating such dialectical constructions of categories of belonging and non-belonging (Lu and Fine 1996: 536; cf. Goody 1982: 26; Bell and Valentine 1997: 168; Wilk 1999: 244; Avieli 2005: 168; Di Giovine 2010). One need only to open a contemporary cookbook, community outreach brochure, or a tourist guidebook to see how food is used to represent and distinguish a particular outsider group or people, often through discourses of heritage and tradition. While there are always variations in any group's cuisine, these often present both a monolithic and unproblematized conception of heritage and food; to wit, the titles of some English-language cookbooks include *Elodia Rigante's Italian Immigrant Cooking* (Rigante 1995), *Authentic Mexican: Regional Cooking from the Heart of Mexico* (Bayless 1987), *The California Heritage Cookbook* (Junior League of Pasadena 1976), and *Mary Sia's Classic Chinese Cookbook* (Sia 2013), but to name a few. The portrayal of ethnicity through these cookbooks can then contribute to the creation and subsequent circulation of ethnic imaginaries inside and outside the group. Garcia-Fuentes et al. demonstrate this in their contribution to this volume, wherein they trace the stunning rise in food-based articles in the Spanish newspaper *La Vanguardia* after the 1977 publication of the seminal *L'art del menjar a Catalunya* (*The Art of Eating in Catalonia*), by popular Catalan mystery writer Antonio Vázquez Montalbán. By 1983, they argue, the recipes depicted in Vasquez Montalbán's book—many of which, by the author's own estimation, had fallen into disuse—were revitalized and unproblematically put into practice as staples of the Catalan diet. This also served to create interest, on the part of locals and Barcelona's government, in restoring and revitalizing the city's traditional food markets, which, too, had fallen into disuse as major national and international supermarket chains gained popularity.

In this way, such culinary media seek to manage and systematize alterity by creating or playing off of stereotypes that are readily comprehendible to their audiences through idioms of food (Appadurai 1998), its consumption remaining one of the most common elements human beings share. They are, furthermore, more than simply representations; they have strong material effects that directly and indirectly shape an ethnic group. Indeed, food has been so linked to the construction of identity that immigration historian Donna Gabaccia (1998) enshrines the old adage, "you are what you eat," into a social fact. This is with good reason. On the one hand, food biologically becomes part of us as we consume and digest it, breaking down into vitamins and minerals that are absorbed by, and circulate throughout, our bodies—catalyzing into the energy needed to live and carry out our individual lives. On the other hand, so-called ethnic food often has distinctive tastes, textures and smells that set it apart from that of the majority; as Bourdieu famously revealed, the cultivation of preferences (or "taste") is itself integral for denoting the "authenticity" of one's membership in such groups (1984: 68). Food thus becomes "readily recognized markers of ethnicity and [perceived] as a major form of traditional culture" (Lu and Fine 1996: 536), a powerful device for the articulation and negotiation of individual and group identity (Wilk 1999: 244).

The chapters in this volume therefore first problematize the very notion of what constitutes an "ethnic" or social group by examining different levels of society for which food-based discourses and practices are utilized to construct identity claims against others. For example, at first glance, Williams-Forson's very poignant and visceral examination of African American foodways seems to deal with a more traditional notion of an ethnic group. However, it quickly becomes evident that inside members of this group entertain radically different understandings, tastes, and imaginaries of what types of food should stand as a marker of African American heritage. Importantly, they break into regional sub-groups. "Soul food"—that seemingly ubiquitously used phrase to denote "down-home" African American cooking in the United States—differs greatly from place to place and even family to family; while creole cookery is marked out as a separate—and at times separately valued—type of cuisine from New Orleans. One of the markers of creole cuisine is that it has very different historical, cultural, and geographic influences, but it is just as true that the regional and even familial variations of "soul food" have developed in their own historical, cultural, and geographic ways as well. Yet varieties in the latter seem to be masked or downplayed in the greater collective American imaginary; creole denotes an authentic, stand-alone cuisine, while different preparations of "soul food" remain simply variations—and, as Williams-Forson reveals, is largely unknown to others within the group. Always lurking behind these identity markers, then, is the issue of value: the choice of how we imagine, and separate out, different types of cuisines even within one ethnic group varies according to how we regard food as a marker of class, authenticity, and heritage.

Just as Williams-Forson shows how New Orleans, or Louisiana more broadly, claims a particular variation of ethnic cuisine as distinct, so too do Heather

Paxson, Warren Belasco, Greg de Saint Maurice, Michael A. Di Giovine, Josep-Maria Garcia-Fuentes et al., Cristina Grasseni, and Carole Counihan demonstrate in this volume that different cities, provinces, or regions within a particular nation-state utilize elements of their foodways to distinguish themselves from their neighbors—problematizing, in the process, the constructed notion of a "national" or otherwise "ethnic" food. Garcia-Fuentes et al., for example, discuss how the revitalization of urban markets in Barcelona is connected to a wider movement to demarcate a regional "Catalan" culinary identity within the Spanish nation-state. This is important, for, as Regina Bendix, Susan Terrio, Miha Kozorog, Erik Castellanos and Sara Bergstresser, and Clare Sammells (all this volume) show, discourses of culinary heritage are often utilized to construct the "imagined community" (Anderson 1983) of the nation-state. Like ethnic food more generally, both groups of contributors reveal the imaginary, or otherwise discursive, construction of identity based on the consumption of food. Indeed, the chapters by Kozorog and Sammells make this particularly evident; Korozog found no clear alimentary basis—other than folkloric—for Slovenia's "national" drink (or drug), the illusive and supposedly hallucinogenic Salamander Brandy, while Sammells shows that the now-ubiquitous llama meat found in curries in Peruvian tourist centers emerged to satisfy touristic imaginaries espoused by Westerners, who strongly associate the country with these animals, and believe consuming llama would be a marker of an authentic tourism experience. If many of these imaginaries trace their origins to rather informal networks of tourists and local entrepreneurs (see also Brulotte and Starkman, this volume), many others are produced within a more formal nexus wherein groups of stakeholders explicitly engage regional and national governments normatively, as Terrio, Bendix, and Grasseni's contributions in particular show. Yet this formal/informal dichotomy is not clear-cut, and all of these contributions in some way reveal that such imaginaries are created through what Di Giovine (2014) has elsewhere called an "imaginaire dialectic"—a complex, formal and informal process of "presenting, imagining, re-presenting, and re-imagining" by groups of stakeholders in a field of production (Bourdieu 1993).

While for African Americans these food-based heritage distinctions rest on social and historical variations, for the places in the United States, Japan and Italy examined by the former set of contributors, an alternative conceptualization of heritage is utilized: that of *terroir*. Often defined as the "taste of a place" (Trubek 2008), *terroir* explicitly links heritage claims with a specific environment, in all of its socio-cultural and natural particularities. This is a very food-specific notion of heritage—one that many different heritage discourses attempt, but which cannot easily be proven—for it posits that the biological components of an environment bestows uniqueness and authenticity onto a place's product, prohibiting it from being replicated elsewhere in the world. That is, a particular heritage food is believed to be authentic—and can be claimed as one group's heritage over another—because of the specific minerals in the soil, weather conditions, and topography, as well as the traditional ways humans have manipulated those environmental conditions. It is believed, and largely verifiable, that these components create a specific physiological "taste" in the ingredients or the dishes that inevitably would change if the same plants or animals were cultivated elsewhere. One of Di Giovine's informants, quoted in his chapter in this volume, explains that while Italy faces real competition by China in replicating its leatherwear and other goods at lower costs, Italy's alimentary products—wine, cheese, meats and vegetables—cannot be reproduced because they are cultivated in the unique terrain and climate that simply cannot simply be replicated elsewhere. In this informant's perspective, the alimentary industry is therefore where the new wave of Italian entrepreneurs should focus their efforts. To wit, in presenting a book on its product, the Consortium for Prosciutto di San Daniele—a form of cured ham that is produced in the autonomous Alpine region of Friuli and a major export to the United States that costs less than the venerable prosciutto di Parma—evocatively constructs *terroir*-based heritage claims thusly:

> The prosciutto di San Daniele is produced, even today, only and exclusively within the confines of the town of San Daniele del Friuli. In the heart of the province of Friuli. And more precisely, on the hills of drumlin [*anfiteatro morenico*], which is a particular system of hills and elevated glacial ridges situated midway between the Alps and the Adriatic Sea, just grazing the Tagliamento, the quintessential Alpine river, where the opposing winds from the North and the South mix, giving life to a particular microclimate. Nature united with the experience of the people of our land—in a spontaneous, natural mixture of knowing and doing—has created a masterpiece of taste and quality. And it is here where the Consortium of Prosciutto di San Daniele has worked for more than forty years to valorize and safeguard this precious heritage. (Consortium Prosciutto di San Daniele n.d.)

This phenomenon is not relegated solely to Italy, or Europe in general. In his contribution to this volume, De Saint Maurice, likewise, examines the case of Kyoto's heirloom vegetables, *kyōyasai*, which partake in the aura of the ancient imperial city as a perceived place of authentic Japanese culture and history. Importantly, in both of these nation-states, there is an obsession with historicity and tracing an authentic foodstuff's "lineage"—or its "provenience," to use a museological term applied to the heritage industry. Indeed, just as heritage is a discourse that links past, present and future, the same can be asserted for *terroir* designations. Yet as Warren Belasco's contribution shows, and as Barbara Kirshenblatt-Gimblett elsewhere intimates (2004a: xii), it is not authenticity *per se* but the "question of authenticity" that creates *terroir* and heritage designations, for through acts of identification, debate, contestation and negotiation, individuals and groups are bound together in a substantive and organized fashion, and the notion of *terroir* takes discursive form.

Of course, *terroir* designations have emerged precisely because of economic concerns, as many of the chapters in this volume make clear. Foods can, indeed, be reproduced outside of their original environments; humanity has a long history of transplanting, replanting, and repopulating their fields and livestock as diverse groups migrate and come into contact with one another, often through colonial encounters (see for example Mintz 1986, Dietler 2010). Yet when conceptualized as cultural heritage, food also has the tendency to adapt and "naturalize" quite quickly; Italians are known for, and indeed, claim, tomatoes as a central component to their cuisine and count them among their primary agricultural exports, and the Irish are likewise known for potatoes; yet both were New World ingredients that were only introduced in Europe during colonial expansion (Coe 1994). There runs the risk, then, that a so-called "indigenous" ingredient can become more profitable, both economically and socio-culturally, in a new locale, among new groups of people—often to the economic detriment of its area of origin. *Terroir* designations emerged precisely with these social and economic components in mind, and exist first and foremost as a "protected mark." Indeed, the first to use this as a "protected mark" was the French government, which sought to control the production of such economically successful viniculture goods as champagne—bubbly wine that is produced in many parts of the world. Yet it has been successful in blocking this type of wine from being called "champagne" if it is not produced in the Champagne region of France; for example, in Italy it is called *prosecco*, or, more generically, *spumante*, and in California it is "sparkling wine." Other than French chocolate or German bread (Terrio and Bendix in this volume, respectively), this has sparked a number of new *terroir*-based protected statuses for other agricultural products. In Italy alone, prosciutto di Parma, mortadella, Parmigiano-Reggiano cheese, and iconic wines such as Chianti all are protected,[1] and a Neapolitan politician made headlines at the turn of the millennium for successfully bestowing protected status on pizza, which, as a "fast food," has become a staple in many non-Italian diets. This development is not limited to Europe. As Brulotte and Starkman describe in this volume, Mexico, too, has a well-developed system for designating certain agricultural goods (as well as the products of other rural industries such as handicrafts) as *marcas colectivas* (literally "collective brands"). Things such as cotija cheese and the large, toasted Oaxacan tortillas known as *tlyudas* now enjoy protected status. The emerging trend to eat locally—including the locavore, Slow Food and "zero-kilometer" movements, in which fresh, local, heirloom foods are valued over imported or factory-produced foods—also have emerged from the same *terroir* concerns.

Certainly these locality-based movements are reactions to the globalization and so-called McDonaldization of the food industry (Barber 1992, Watson 2006, Wilk 2006), as well as the pirating of food brands (such as Wisconsin-made

---

1   See, for example, the "Legends from Europe" campaign financed by the European Union and the Italian ministry of agricultural policy, which promotes Italian foods that have protected status: www.legendsfromeurope.com (accessed on 16 June 2013).

"Parmigiano-Reggiano"), in Italy as described in this volume by Carole Counihan and Cristina Grasseni, as in other parts of the world. But particularly in Italy, it is also a counter-reaction to the influx of North African and Chinese immigrants who, espousing markedly different physiognomic features, religions and worldviews, and food, have created extreme socio-cultural tensions within the nation-state. As Pagliai (2009) shows in her study of local prejudices towards new immigrants in the Tuscan city of Prato, "ethnic" food—its look, its taste, and especially its smell, wafting out from the homes of these new immigrants—is often the source of complaints by locals in some small cities. Through the materiality of ethnic food and food-based experiences, complex, difficult-to-articulate problems and prejudices can find expression. Indeed, Counihan's study of the Sardinian Slow Food movement in this volume describes how several towns have begun to legislate bans on the opening of "ethnic food" establishments in their historic centers (*centri storichi*) under the guise of keeping their tourist centers "authentic." Beyond xenophobia, and beyond merely economic concerns (though certainly cultural tourism plays a distinctive part in Tuscany and Sardinia—and in the Alpine areas described by Grasseni in this book), what we observe in Italy is the process of establishing new ethnic groups and erecting new boundaries that differentiate a "non-ethnic" majority (the "Italians") from North African and Chinese minorities. This is accomplished through the (pejorative) ascription of food heritage-based stereotypes in much the same way that nineteenth-century Americans did with Italian immigrants (see Cinotto 2013, 2009: 659, Levenstein 1985: 7–12; cf. Covello 1958: 24–5). The Slow Food movement and the establishment of agricultural eco-museums (see Grasseni this volume) are essentially ways of authenticating "true" Italians from ethnic minorities who are not rooted (and perhaps are seen as not belonging) in the land.

But despite the seemingly cut-and-dry normative and ideological components of *terroir* discourses, it is important to note that even these designations are more fluid than they may at first seem. To continue with the Italian example, these same new immigrant groups count among the producers of such "protected" foods as pizza and even Prosciutto di Parma. This complicates not only the notion of what, or who, contributes to creating an authentic "taste of a place" (must socially ascribed "Italians" be the sole producers of these authentic goods?) but it also complexifies performances of ethnic heritage. What stands out in many negative discourses concerning more recent Pakistani and North African immigrants in both informal discussions among Italians and in the Italian news media are their alternative religious values, such as their religious taboo on consuming pork. But particularly in the case of Muslim workers in Parma's prosciutto factories, the elaboration of this pork-based food quite literally challenges their authentic performance as good, halal-observant Muslims. Italian religious attitudes are also reflected in the marketing decisions of some Muslim pizzeria owners, who, perhaps in an attempt to appeal to ethnic Italians and seem more "authentic" (or less exotic), name their restaurants after prominent Catholic saints such as the immensely popular twentieth-cenury Catholic stigmatic and saint, Padre Pio of Pietrelcina (see, for example, Ferraresi 2011). Are these choices to perform Italianness, or at least to transgress Islamic norms, motivated solely by economic concerns, or may we also see this as indexical of a gradual "acculturation" into Italian culture and worldview? Are explicit performative transgressions of minority religious-ethnic norms necessary to authenticate the heritage of the protected "Italian" food they produce?

As Di Giovine, Grasseni, and Counihan discuss in their chapters, there now is an increasing interest by Italian regions and localities, as in other places (see, for a U.S. example, Shortridge 2004), to differentiate themselves by starting food-centered *sagre* (popular festivals), eco-museums, and World Heritage nominations, in an attempt to construct *terroir* designations—if not legally, then at least in the collective psyche. On the other side of the Atlantic, Warren Belasco and Heather Paxson's contributions to this volume reveal that one can cogently construct, and argue for, the application of *terroir* designations for many culinary traditions—even when many of the practitioners themselves do not readily use such idioms. Belasco makes a convincing case for *terroir* in Washington D.C., a U.S. city that is rarely, if ever, considered to possess a geographically rooted cuisine. Paxson in particular goes beyond this when she argues that the defining aspect of a region's heirloom cheese production is less tied to environmental conditions as it is social ones: a "tradition of invention." While playing on Hobsbawm and Ranger's (1983) famous notion of political and ritual processes as an "invention of tradition," Paxson points to what heritage theorists consider "intangible cultural heritage"—a "metacultural production" that consciously delineates the "totality of tradition-based creations of a cultural community" (Kirshenblatt-Gimblett 2004b: 54). In sum, Paxson posits that a region's *terroir* can be linked not exclusively to the environment's biological components, but rather to what Italian economists are calling the "genius loci," or "spirit of a place"—an almost genetic tie to one's land of origin that spawns a particular habitus and processes of production that simply are not replicable elsewhere (see, for example, Cipolla 1990, Balloni and Trupia 2005, Matacena and Del Baldo 2009, Del Baldo and De Martini 2012, cf. Norberg-Schultz 1980). It is this notion of being tied to the land in a socio-biological fashion, which manifests itself in one's food preferences, that is at the heart of many food-based heritage claims.

UNESCO and Food as Intangible Cultural Heritage

In 2010, food made its first appearance on UNESCO World Heritage lists with the inscription of three diverse cuisines and one type of culinary product: the Mediterranean Diet, the French Gastronomic Meal, Mexican/Michoacán Cuisine, and Croatian gingerbread. This is significant, for it underscores the normative centrality of food and food-based imaginaries pertaining to cultural heritage. It is important to understand the development of the Intangible Heritage List because it has had profound effects on the ways in which cuisine has entered into a global political regime that demarcates knowledge and practices worthy of recognition and preservation.

Though the Convention for the Safeguarding of Intangible Cultural Heritage was only ratified in 2003, and its corresponding List was only adopted in 2006, in its current form, it emerged from UNESCO's longer, evolving role in ostensibly promoting international peace. Founded at the end of World War II, the United Nations Educational, Scientific and Cultural Organization was created to complement the more top-down post-war reconstruction and peacekeeping efforts by the Marshall Plan and the United Nations by engaging with publics at the intangible level (see Di Giovine 2013):

> For this specialized UN agency, it is not enough to build classrooms in devastated countries or to publish scientific breakthroughs. Education, science, culture and communication are the means to a far more ambitious goal: to build peace in the minds of men. (UNESCO 2003b: 1)

The Preamble to UNESCO's 1945 Constitution makes it clear that peacemaking begins in the minds of individuals, rather than in the embassies and legislatures of nation-states, and therefore the organization's duty centered on knowledge production and dissemination (see UNESCO 1972: 1), such as assisting in educational endeavors, promoting scientific research and cultural preservation (i.e. an appreciation of cultural diversity, in the broadest sense). "Peace must therefore be founded, if it is not to fail, upon the intellectual and moral solidarity of mankind," the Preamble states unequivocally (UNESCO 1945: 1).

Interest in designating and conserving cultural and natural heritage, however, only became a focal point of UNESCO's work in the aftermath of the organization's successful mobilization of the international community to save the thirteenth-century Egyptian temples of pharaoh Ramses II at Abu Simbel in Nubia from floodwaters of the Aswan High Dam, and to protect Venice from flooding. Both of these initiatives revealed the very affective, communal and mediatory nature of heritage; far from being localized in one group, it was possible for many peoples to identify a monument as part of their heritage, even if historically they had little claim to, or immediate interest in, it. In 1966, UNESCO's then-Director-General Renée Maheu, in fact, underscores that the most important outcome of these "events" was an increase in global awareness that heritage policies, at the political level, could lead to this intangible "intellectual and moral solidarity of mankind" that is at the root of UNESCO's objectives. Meeting to discuss conservation policies in 1972, UNESCO's member-states ratified the World Heritage Convention, which would help to protect, through inscription on a prestigious World Heritage List, sites of "outstanding universal value." It is important to note that a country technically offers up (UNESCO 2008: 5) its heritage property to the international community but receives little to no funds for it; rather, the prestige and awareness raised by being on the list often (but not always) translates into material assistance from donors and conservation groups in the short-term, and is thought to create economic growth through tourism in the medium-term (Di Giovine 2013). Most importantly, claiming that a group's specific heritage is "universally valued" often is perceived as valorizing that particular social group, and integrating it further into the global community (Di Giovine 2009a, 2009b).

As the years progressed, however, it became clear that the World Heritage List gave disproportionate representation to Western countries, since most of the sites considered to be of "outstanding value" were large monumental buildings, religious edifices, and objects that fit particularly Western aesthetic models. Though UNESCO implemented a relatively successful *Global Strategy for a Representative, Balanced and Credible World Heritage List* (1994) to broaden its interpretation of what could be valued, it did little to alleviate tensions among developing nations and aboriginal groups whose cultures were predicated more on immaterial practices than the creation of monumental material cultural forms. UNESCO responded by taking progressive steps towards defining, valorizing and safeguarding intangible cultural heritage. Building on a failed 1989 *Recommendation for the Safeguarding of Folklore and Traditional Culture*, a Korean proposal to create a "World Living Human Treasures List, similar to the World Heritage List" (UNESCO 1993: 2), and a sweeping study of intangible heritage (UNESCO 1996), in 1997 UNESCO adopted a resolution creating the *Proclamation of Masterpieces of the Oral and Intangible Heritage of Humanity*, a type of List that was not, however, subject to a Convention and had no financial resources at its disposal (Hafstein 2009: 95). Though it was intended to "proclaim" the top forms of immaterial heritage, it went largely unnoticed in comparison to the World Heritage List; it was also criticized by many of the same indigenous groups as being too hierarchical and elitist, and without adequate reliance on the participation of the communities producing such masterpieces. While all groups pushed for a Convention that would provide legal instruments similar to that of the World Heritage Convention, they were divided in their desire for a List; one faction felt that a selective List would be the only adequate way of raising the necessary awareness for safeguarding these practices, while the smaller nations wished to eschew lists altogether (or at least a selection process), which they perceived as unconditionally elitist. The final product, *The Representative List of the Intangible Cultural Heritage of Humanity*, and its associated *List of Intangible Cultural Heritage in Need of Urgent Safeguarding* (see UNESCO 2003a), represents a compromise between the two: while still selective in its process of inscription, it eschews "universal value" as its primary criterion in favor of "representivity" (Hafstein 2009: 108); each signatory draws up a complete inventory of its cultural practices, and is encouraged to submit one for inscription during the Intangible Heritage Committee's biennial meeting. Importantly, as Kirshenblatt-Gimblett points out, it "entails a shift from artefacts (tales, songs, customs) to people (performance, artisans, healers), their knowledge and skills" (2004b: 53). People, knowledge and skills are things that all groups possess, though the material outcomes of their interface may differ substantially. It thus builds on the basic structure of the World Heritage List, yet purports to be more democratizing and even less determinate in its standards of selection—thus attempting to preclude a situation of unbalanced representationality that the Global Strategy was forced to correct (Di Giovine 2013).

The Representative List of the Intangible Cultural Heritage of Humanity seems to be a key development for the evolution and survival of the World Heritage List. While separate, both the tangible and intangible Lists are inexorably linked historically, conceptually, and, to a certain extent, even procedurally. One was born from the other so that the other could remain relevant amid changing conceptions of heritage (Di Giovine 2013). Indeed, the World Heritage List has progressively moved from designating sites primarily conceived as artifactual and monumental, to include more holistic, living notions of heritage. Importantly, intangible heritage designations have quickly been met with similar celebrations at the local level in many societies across the world, and in its early years, locals often conflated the Intangible Heritage List with that of the World Heritage List. Particularly following the Yamato Declaration, in which participants agreed to create a more integrated approach towards safeguarding both tangible and intangible heritage (UNESCO 2004), it is conceivable that the Intangible Heritage list, while remaining separate and decidedly less place-oriented, will implicitly inform future nominations by states-parties, as well as future actions by the World Heritage Convention.

Almost immediately after the ratification of the Intangible Heritage Convention, nation-states in Latin America and Western Europe, for whom food already factored into heritage claims and touristic imaginaries, began drawing up inventories cataloguing, and thereby constructing, a systematic narrative about their cuisines. While these were political processes in and of themselves, as Regina Bendix makes clear in her contribution, they also aimed to engage different forms of culinary "experts" to construct such lists; indeed, elsewhere Bendix, Eggert and Peselmann have emphasized the power of "go-betweens" and "interpreters" for constructing, and diversifying, what they call UNESCO's "heritage regime" (2012: 19–20). Mediterranean countries, including Italy and France, for example, employed academic anthropologists trained in folklore to conduct participant observation, surveys, and interviews with restaurateurs, agriculturalists, and citizens in designated areas (Bortolotto 2012, Broccolini 2012, Tornatore 2012), while Mexico employed the aid of internationally renowned chefs and state tourism promoters (see Sammells' and Brulotte and Starkman's contributions in this volume). This has led to different conceptualizations of whose food and cuisine was worth elevating to the status of national (or international) importance, what aspects of such foodways was of value, at which time periods and how long must such food-based practices be enacted before they can be considered "heritage," how the international community should value the food, and even how one goes about identifying and promoting it.

The outcomes were radically different; one was designated a "diet," another was a "meal," another a "cuisine," and another was a particular dish.

Italy, Spain, Morocco and Greece[2] produced a trans-national designation of the "Mediterranean diet" which focused on culinary commonalities within the Mediterranean basin that had already been vaguely known in Western circles as the basis for good nutrition and longevity;[3] they identified the longstanding cultivation and consumption of olive oil, fresh fruit and vegetables, seafood, and certain grains, and wine—as well as the skillsets, folkloric knowledge, and attention to custom and tradition "ranging from the landscape to the table" that has made the Mediterranean Diet known (or rather, perceived) outside of those countries as "a nutritional model that has remained constant over time and space" (UNESCO 2010c). Of course, as Garcia-Fuentes et al. discuss in this volume, and Helstosky in her study of food and politics in Italy asserts, imaginaries concerning an unchanging and traditional diet of Mediterranean peoples was a conscious product on the part of national and regional stakeholders to combat "the challenge of abundance" and changing food consumption trends in the area (2004: 127–8).

While the Mediterranean Diet was packaged as a broad and seemingly unchanging "nutritional model" of many peoples with a long history of social and environmental interaction, Mexico did the inverse: while still emphasizing longstanding tradition, ritual and folkloric elements, it designated a particular regional cuisine, from the state of Michoacán, that could stand in as a metonym for the culinary heritage of the country as a whole, despite a diversity of its people and its food-based traditions. Entitled "Traditional Mexican Cuisine—Ancestral, Ongoing Community Culture: The Michoacán Paradigm" (UNESCO 2010d), the listing emphasized similar folkloric qualities and skillsets as the Mediterranean diet, as well as common alimentary elements such as beans, maize (corn) and chile (peppers) that are perceived to have a long history in the country, but it specifies a particular regional instantiation of these components as metonymic for the culinary heritage of the country as a whole. As Sammells intimates in this volume, just as the Mediterranean diet had already entered the global imaginary as a form of heritage cuisine, Michoacán cuisine, too, had already been understood and represented abroad as "Mexican" cooking. Again, the involvement of intermediaries were important: while employing social scientists to construct a compelling inventory of alimentary elements, the Mexican state also utilized the expertise, as well as the prestige, of internationally renowned (and internationally educated) "top chefs" to both construct and promote the designation. Yet as Brulotte and Starkman reveal in this volume, even Mexican entrepreneurs promoting their indigenous Oaxacan soup, *caldo de piedra*,

---

2   In 2013, the Mediterranean Diet was re-inscribed on the List with an expanded group of nation-states that also included Croatia, Cypress and Portugal (UNESCO 2013a).

3   In a brief open-ended survey, Di Giovine asked 100 respondents to list what they knew of the Mediterranean Diet; many of these elements, particular olive oil, fish, grains and wine were commonly listed. For more information on the survey, see Di Giovine forthcoming.

conflate these categories, purposefully advertising their regional Oaxacan (and not Michoacán) cuisine as "intangible cultural heritage."

Croatia, in its designation of "Gingerbread Craft from Northern Croatia" (UNESCO 2010b), took another complementary route: it designated as distinctively "Croatian" a type of food that is at once transnational and transhistorical, as well as regional ("Northern Croatian"), and made it uniquely its own. Its designation narrative is telling:

> The tradition of gingerbread making appeared in certain European monasteries during the Middle Ages and came to Croatia where it became a craft. Gingerbread craftspeople, who also made honey and candles, worked in the area of Northern Croatia ... The craft has been passed on from one generation to another for centuries, initially to men, but now to both men and women. Gingerbread has become one of the most recognizable symbols of Croatian identity. Today, gingerbread makers are essential participants in local festivities, events and gatherings, providing the local people with a sense of identity and continuity. (UNESCO 2010b)

While in the Mediterranean Diet and "Mexican" cuisine, a narrative is crafted that pinpoints its value to the geographic area located within the nation-state claiming it, as well as to the deep past, Croatia makes the value-claim that, while gingerbread production was—and still is—practiced by a variety of peoples, it was perfected into a "craft" by Croatians, and has become a central component in Croatian culture.

Lastly, the much-lauded (in French media) "Gastronomic Meal of the French" further expands Croatia's tactic. Rather than designating a particular cuisine, particular forms of "traditional" cultivation, or even particular types of foods and ingredients, this designation focuses on what is often perceived in culinary imaginaries as a quintessentially French (or generally Mediterranean) ways of consumption. Importantly, these elements are extremely vague, and can be extrapolated to many other cultures throughout time and space, and it is worth citing the whole of UNESCO's description in illustrating this point:

> The gastronomic meal of the French is a customary social practice for celebrating important moments in the lives of individuals and groups, such as births, weddings, birthdays, anniversaries, achievements and reunions. It is a festive meal bringing people together for an occasion to enjoy the art of good eating and drinking. The gastronomic meal emphasizes togetherness, the pleasure of taste, and the balance between human beings and the products of nature. Important elements include the careful selection of dishes from a constantly growing repertoire of recipes; the purchase of good, preferably local products whose flavours go well together; the pairing of food with wine; the setting of a beautiful table; and specific actions during consumption, such as smelling and tasting items at the table. The gastronomic meal should respect a fixed structure, commencing with an *apéritif*

(drinks before the meal) and ending with liqueurs, containing in between at least four successive courses, namely a starter, fish and/or meat with vegetables, cheese and dessert. Individuals called gastronomes who possess deep knowledge of the tradition and preserve its memory watch over the living practice of the rites, thus contributing to their oral and/or written transmission, in particular to younger generations. The gastronomic meal draws circles of family and friends closer together and, more generally, strengthens social ties. (UNESCO 2010a)

Such is the content expressed in the expertly crafted video produced to sway voting members of the Intangible Heritage Committee.[4] Its images show a conglomeration of foods being bought, sold, and eaten, without providing any precise names; it shows different peoples (including those of different racial origins) talking together, purchasing and cooking food together, consulting with cooks and shop-owners, and discussing their food-based meals with one another. It posits that the food-based heritage of the French is an innately lively, group-centered affair. Important is the element of time: While the video produced for the Mediterranean Diet[5] emphasizes long, languid pans over lush olive groves and scintillating azure seas, and features fishermen and farmers sometimes in traditional "peasant" dress, the French video eschews imaginaries of the unchanging and natural past for contemporary life, and contemporary social interactions that are as varied as the peoples who constitute French society. Indeed, a central premise of this volume—a truism within the social scientific study of food—is that the consumption of food is a practice in commensality; it brings people together in an often structured or ritualized way, it emphasizes in both discourse and practice the aspect of togetherness, of social exchange, and often is made to elicit precise memories of people, places and events in the past. In the words of famed food writer M.F.K. Fisher, "There is communion of more than our bodies when bread is broken and wine drunk" (quoted in Counihan and Van Esterik 2008: xi). Yet France's narrative does far more than simply claim possession, or patrimony, over a cultural practice shared by many peoples; it underscores in the most complete and paradigmatic way the very nature of heritage discourse itself: heritage is a claim that ultimately mediates between individuals within a social group, strengthening social ties within the society spread out over time and space, imaginarily linking past and present. There is, indeed, a certain nostalgia imbedded in much of heritage discourse—an understanding that certain elements of the past, be they tangible or intangible, are worth preserving and conveying to future generations. Particularly in light of recent scholarship describing the current era as a "post-traditional age," in which rituals are perceived to have lost their relevance (D'Agostino and Vespasiano 2000: 5),

---

4 The video may be found on UNESCO's website: http://www.unesco.org/culture/ich/RL/00437.

5 The video for the Mediterranean Diet may also be found on the UNESCO website: http://www.unesco.org/culture/ich/RL/00884.

heritage discourses and practices stand in stark contrast, complicating this notion. Yet the Gastronomic Meal of the French also, importantly, reveals the changing nature of heritage—something that goes against many of the guiding principles of UNESCO's original World Heritage Convention. While UNESCO, and indeed most heritage practitioners and preservationists, have been accused of privileging a sort of "museumification" of cultural practices—as they attempt to stave off the inevitable transience of time (Di Giovine 2009: 81, 301–27) or at least attempt to restore institutions and practices that seem to have lost their relevance in the contemporary era as Bendix's breadmakers and Terrio's chocolatiers (in this volume) seem to have—the Gastronomic Meal of the French builds change directly into the designation by identifying the vaguest of practices and the most general of concepts as heritage. This should be seen neither as a weakness to the designation nor as a cop-out; rather, we argue, this may well represent the changing nature of heritage itself—particularly where the ongoing, universal consumption of food is concerned.

Indeed, at the end of 2013 Japan succeeded in designating their cuisine, *washoku*, as Intangible Cultural Heritage (UNESCO 2013b). Building on the holistic approaches of the Mediterranean Diet and the French Gastronomic Meal, as well as on their emphases on healthy eating and well-being, "Washoku, traditional dietary cultures of the Japanese, notably for the celebration of New Year" seems to self-consciously imitate the French tactic:[6] rather than discussing a particular food or set of foods (like Croatian gingerbread or Turkish coffee, also designated in 2013), or being relegated to a distinct region or area (such as Michoacán or Mediterranean cuisine), or even focusing on the immutable past, the Japanese designation—like the French—emphasizes the way in which shared cultural values towards the social and natural world are played out and made manifest through the preparation and consumption of food, and what is to be passed down from generation to generation.

Conclusion: Revitalizing Economies, Remaking Identities

That the World Heritage List is a vehicle for economic development is well established. Yet it does not necessarily take inscription on this global inventory to draw attention to so-called heritage foods' economic potential. Di Giovine's chapter in particular illustrates the way in which a small town's local food, previously considered *cucina casareccia*, or peasant cuisine,[7] has, with the help of pilgrims and outsiders visiting the village, transformed itself into highly valued

---

6   We would like to thank Ted Bestor for pointing this out in his paper on the Japanese nomination during the session, "Edible Identities," at the American Anthropological Association's 2013 conference, at which several of the contributors to this volume spoke.

7   Literally "home cooking," but in this area especially, it has the connotation of a rustic, agrarian cuisine (also known as "*cucina povera*").

"heritage cuisine" that is able to convey to these outsiders a distinctive, taste-based experience of the place, its people, and its favorite son, St. Padre Pio of Pietrelcina.

In a way it resembles the cycle of rebirth that Penny Van Esterik (2006) describes in her exploration of the way in which Lao "hunger food" is converted into economically valuable "heritage food"—food associated with poverty and the lower classes at the local level is picked up by cosmopolitan outsiders, particularly those involved in cultural and culinary tourism—as a marker of exoticism, but with notable exceptions. First, Van Esterik's "hunger foods" are those elements (bugs, algae) that are foraged to stave off periods of extreme hunger, and in that way are marked as extra-ordinary or non-quotidian dishes; similar to William-Forson's "soul food," the food discussed by Di Giovine in this volume are unmarked by locals; they are relatively quotidian—a type of "down home cooking" that is (believed, at least) to be common at the tables of their compatriots. Second, there is a clearer binary in Van Esterik's paradigm than in Di Giovine's: what is typically hunger food for Lao are heritage food for non-Lao (primarily Westerners). In Di Giovine's case of Pietrelcina, however, a hierarchy is intimated: Pietrelcinese consider artichokes and ciciatelli pasta to be "theirs," a unique form of "ethnic" cuisine that marks them out from others. But the meanings are different for outsiders. On the one hand, non-Italians such as the Irish and the Americans seem to perceive this pasta more broadly as a "taste of Italy," some Italian food to bring home from their trip; on the other hand, other Italians—the primary consumers of these goods—buy these pastas in bulk because they recognize them as being not "Italian" but "Pietrelcinese," available nowhere else but here. For all parties, however, these pastas are imbued with a narrative that links them with the saint: by consuming them, they are acting out not their cosmopolitan dispositions towards the exotic, but rather attempting to commune with the saint through a visceral experience with place. Third, Van Esterik's heritage food was valuable mainly because of its exoticism and difficult-to-obtain nature outside of the particular location; one had to make the costly and time-consuming trip to the small, insular Southeast Asian country to truly experience it, or else they had to find an importer to obtain them in their home countries. However, pasta itself is not difficult to find in the United States and Ireland (for the latter, see D'Aurea 2010); rather, these can be thought of as "edible souvenirs" that, when consumed, could spark memories of their time on the pilgrimage—and even help them "pray better" (cf. Di Giovine 2012: 116).

Indeed, as Sammells' and Di Giovine's chapters show, culinary heritage and travel are often linked, even in discourse; cookbooks often contextualize the cuisine by setting the stage, impressing upon its readers to put themselves in the shoes of the tradition-bearers, either by recounting stories of immigration or asking readers to imagine that they are on an oft-exotic voyage to the cuisine's "homeland." Tourism is an engagement with alterity (Picard and Di Giovine 2014), as is sampling or otherwise (re-)presenting a "foreign" or different ethnic cuisine (cf. Long 2004). Most importantly, both are sustained by "imagineering otherness" (cf. Salazar 2013). It is also important to note that, with the exception of community-produced brochures perhaps, many of these culinary media are penned not by insiders but outsiders themselves who, through such accessible narratives, as well as through modifying their recipes, often "tame" or "acculturate" the food itself to the palates and sensibilities of their readers. In the end, however, "to write about food is to write about the self," insists Goldman (1992: 169), and as the anthropological truism goes, we study (and write about) cultural Others to learn about ourselves as well (cf. Crick 1995). Cookbooks, culinary travel writing, and food-based heritage claims are salient, therefore, because they are geared not only (if at all) to ethnic insiders who "know" how to produce a particular cuisine, but also to outsiders as a process of representation and integration (Appadurai 1988). They are manuals not only for the production of an ethnic cuisine, then, but for the production of ethnic imaginaries *par excellence*.

In the end, however, these media go beyond mere discourse; the recipes they contain are meant to be put into practice, both through cooking and, more importantly, through consuming. On the one hand, the recipes are meant to be cooked; they allow us to get a glimpse of another's heritage by asking us to replicate "traditional" processes, and use sometimes "traditional" or uncommon ingredients. As Lisa Heldke (2001) mused during a Thailand-inspired dinner party, we frequently do not open an ethnic cookbook and think about replicating a recipe for a particular dish like Panang chicken, but we say, "Let's cook Thai." In those simple words, we invoke distinctive heritage claims: this dish is not an ordinary dish; it is not "my" dish, but a Thai dish—a dish that is part of the patrimony of some other group; that which signifies the Other, is owned by the other, and passed down to generations of Others. Yet in that act of "cooking Thai," or cooking Italian, Mexican, or Chinese, we share in that privileged patrimony. However, Heldke (2003) offers an even more critical take on these acts of culinary translation by pointing out how they are often embedded in larger historical projects of colonialism and empire. For instance, the seemingly harmless impulse of Anglos in the U.S. to cook or eat "Mexican"—and the various industries it has spawned, from upscale restaurants to chef Rick Bayless's PBS television show, *Mexico: One Plate at a Time*—does not acknowledge the complicated, often fraught history of social and economic relations between the two countries. In light of ongoing anti-immigrant sentiment, Americans' embrace of "Mexican food" (shorthand for what are in reality highly varied, regional cuisines) apparently does not uniformly extend to the groups from which it emanates—an irony not lost on Gustavo Arrellano, a Mexican-American journalist and writer who has documented the rise of Mexican food in the United States (see Arrellano 2012).

At the same time, when members of an ethnic group mark their cuisine as "ethnic"—particularly during festivals or ritualized meals, such as Sunday pasta dinners or the Christmas Eve "Feast of the Seven Fishes" for Italian-Americans (Di Giovine 2010)—the heritage component takes on added salience as a means of re-asserting, and even revitalizing, ethnicity itself. This is particularly important for negotiating second- and third-generations' "hyphenated identities" (Trinh 1991), both through cooking and, more importantly, through consuming. "Ethnic" restaurants, like cookbooks, offer a "taste of" the country they represent, regardless of whether it is a corner *taquería* (taco stand) run by immigrants, a high-end restaurant with professionally trained chefs, a fast-food chain like McDonalds offering breakfast burritos, or a chain restaurant like the Olive Garden, which constructs its authenticity by invoking stereotypes of large Italian feasts and advertising its own cooking school, the Culinary Institute of Tuscany (see Olive Garden 2014).

While perhaps an unusual step for a commercial chain restaurant in the United States, the fact that the Olive Garden has opened a cooking school in Tuscany underscores the importance of food in forms of touristic encounters as well—something many, if not all, of these chapters implicitly recognize. The contributions by Belasco, Brulotte and Starkman, Counihan, Grasseni, Sammells, Di Giovine, Korozog, and Garcia-Fuentes et al., all reveal a real concern, on the part of different groups of stakeholders, with the ways in which outside visitors experience a site—and locals' heritage—through their food. With its Culinary Institute, the Olive Garden has capitalized on the burgeoning culinary tourism industry, as well as the dominant "tourist imaginaries" (see Salazar and Graburn 2014) of Italy as centered on food; endorsements from traveler-participants on its webpage also attest to this. To wit, a traveler named Brandon Wicks is quoted on their website as saying, "The Italian people have a love for food. They don't just eat food to get energy for the body, they use all their senses. They love the food, they smell it, they taste it, they sit at a table and build connections with other people ... the culture is just what the Olive Garden describes" (Olive Garden 2014). Yet what is interesting about food and tourism is that one need not travel extreme distances to be a "culinary tourist"—and indeed, as Cohen and Avieli (2004) point out, it is often more difficult to engage in than what promotional media often portray. One can engage in culinary tourism simply by visiting an ethnic restaurant in their own city—including the Olive Garden, a chain that is found in most U.S. urban areas (see Albrecht 2011). Indeed, in periods and in places where travel was restricted or otherwise impossible, these restaurants were simultaneously a way to escape the realities of political oppression and immobility, and to mark a certain level of cosmopolitanism (Caldwell 2006). Either way, like other forms of tourism (Graburn 1983), culinary tourism—whether domestic or foreign—is a way to escape the travails of everyday life, often serving as a revitalizing "rite of intensification" (Chapple and Coon 1942: 398–426) that refreshes the social order, as well as the participant's sense of well-being.

We conclude by suggesting that a foodway itself is marked by what may be considered multiple iterations of revitalization. All of the essays in this volume address, in one manner or another, how particular foods and associated practices are born, reborn, and reborn again as different types of components in a food chain. That is, each cycle of rebirth marks a different significance of food: from plant or animal (cow or palm tree) to commodified object to ingredient that is elaborated upon (beef or coconut) to a menu item or "dish" (hanger steak or curry); when coupled with other dishes, furthermore, a "dish" gains new meaning and new value as a cuisine ("American cuisine" or "Thai cuisine"). In examining the ways in which these foods impact, and are impacted by, social situations, food is reborn again and again; as an offering, a source around which commensality exists or gastropolitics are played out, as markers of socio-economic distinctions such as age, gender, kinship, or class, and importantly as valuable sources of a group's identity or heritage that connects its individual members to each other through time and space.

In the end, what the volume as a whole intends to show, and what we have argued here, is that when conceiving of food as heritage—while closely implicated in what is sometimes uncharitably called the "heritage industry" (Hewison 1987) and thus is subject to commodification and standardization (see Comaroff and Comaroff 2009)—we should also examine the strong socio-cultural impacts that often are often subtle, viscerally experienced on individual levels, and subject to long-term processes of transformation. These are what we have called here "edible identities."

Acknowledgements

This book could not have been produced without the support of a number of individuals and institutions. The idea of this volume was born out of a very productive conversation with Dymphna Evans, Publisher at Ashgate, during the first international congress of the UNESCO/UNITWIN Network "Culture, Tourism, Development," held in Quebec City in June 2010. We thank the organizers for a provocative meeting that set the stage for this collaboration, and we are especially grateful for the support of our editors at Ashgate throughout this process. We also acknowledge the hard work of our contributors to this volume, who enthusiastically took up our invitation. Much of the theory that guided the theoretical trajectory of this volume emerged through discussions and presentations at a number of conferences, and we especially thank Ted Bestor and Michael Herzfeld for their comments during a double panel session at the American Anthropological Association's annual meeting in Chicago in 2013, at which several of these papers—as well as a working draft of this introduction—were presented. The work invested in the writing of this introduction and the editing of this volume was supported, in part, through funds from the Hanna Holborn Gray-Andrew Mellon Fellowship for the Humanities and the Humanistic Social Sciences at the University of Chicago, the Department of Anthropology and the Latin American and Iberian Institute at the University of New Mexico, the Department of Anthropology at the University of Wisconsin-Madison, and an AWA grant from the Department of Anthropology and Sociology at West Chester University of Pennsylvania. We would especially like to thank Shane Metivier and Abigail O'Brien at West Chester University for their assistance in compiling the index to this volume.

References

Alba, R. 1990. *Ethnic Identity: The Transformation of White America.* New Haven: Yale University Press.

Albrecht, M. 2011. "When You're Here, You're Family": Culinary Tourism and the Olive Garden Restaurant. *Tourist Studies* 11(2): 99–113.

Anderson, B. 1983. *Imagined Communities: Reflections on the Origin and Spread of Nationalism.* London: Verso.

Appaduri, A. 1988. How to Make a National Cuisine: Cookbooks in Contemporary India. *Comparative Studies in Society and History* 30(1): 3–24.

Arrellano, G. 2012. *Taco USA: How Mexican Food Conquered America.* New York: Scribner.

Asad, T. 1993. *Genealogies of Religion.* Baltimore: Johns Hopkins Press.

Avieli, N. 2005. New Year Rice Cakes: Iconic Festive Dishes and Contested National Identity. *Ethnology* 44 (2), Spring: 167–87.

Bakalian, A. 1993. *Armenian-Americans: From Being to Feeling Armenian.* New Brunswick, NJ: Transaction Books

Balloni, V. and P. Trupia (eds) (2005), *Origine, caratteristiche e sviluppo dell'imprenditorialità nelle Valli dell'Esino e del Misa*, Ancona: Conerografia.

Barber, B. 1992. Jihad vs. McWorld. *Atlantic Monthly.* March. http://www. theatlantic.com/magazine/archive/1992/03/jihad-vs-mcworld/303882/ (Accessed on 11 January 2014.)

Barth, F. 1969. *Ethnic Groups and Boundaries: The Social Organization of Cultural Difference.* Boston: Little, Brown and Company.

Bashkow, I. 2004. A Neo-Boasian Conception of Cultural Boundaries. *American Anthropologist* 106(3): 443–58.

Bayless, R. 1987. *Authentic Mexican: Regional Cooking from the Heart of Mexico.* New York: William Morrow.

Bell, D. and G. Valentine 1997. *Consuming Geographies: We Are Where We Eat.* New York: Routledge.

Bendix, R., A. Eggert and A. Peselmann (eds) 2012. *Heritage Regimes and the State.* Göttingen: Universitätsverlag Göttingen.

Bortolotto, C. 2012. The French Inventory of Intangible Cultural Heritage: Domesticating a Global Paradigm into French Heritage Regime. In R. Bendix, A. Eggert and A. Peselmann (eds), *Heritage Regimes and the State.* Göttingen: Universitätsverlag Göttingen, pp. 265–82.

Bourdieu, P. 1984. *Distinction: A Social Critique of the Judgment of Taste.* Trans. Richard Nice. Cambridge, MA: Harvard University Press.

Broccolini, A. 2012. Intangible Cultural Heritage Scenarios within the Bureaucratic Italian State. In R. Bendix, A. Eggert and A. Peselmann (eds), *Heritage Regimes and the State.* Göttingen: Universitätsverlag Göttingen, pp. 283–301.

Chapple, E. and C.S. Coon 1942. *Principles of Anthropology.* New York: Holt.

Cinotto, S. 2009. La Cucina Diasporica: Il Cibo come Identità Culturale. *Storia d'Italia.* Annali 24 Migrazioni: 653–72.

Cinotto, S. 2013. *The Italian American Table: Food, Family, and Community in New York City.* Urbana: University of Illinois Press.

Cipolla, C.M. 1990, *Storia economica dell'Europa pre-industriale.* Il Mulino, Bologna.

Coe, S.D. 1994. *America's First Cuisines.* Austin: University of Texas Press.

Cohen, E. and N. Avieli 2004. Food and Tourism: Attraction and Impediment. *Annals of Tourism Research* 31(4): 755–78.

Comaroff, J. and J. Comaroff 2009. *Ethnicity, Inc.* Chicago: University of Chicago Press.

Consortium Prosciutto di San Daniele. n.d. "San Daniele: Natura, Gente, Città, Prosciutto dei Friuli." http://www.prosciuttosandaniele.it/home_prosciuttosandaniele.php?n=160&l=it (Accessed on January 16, 2014.)

Counihan, C. and P. Van Esterik. 2008. *Food and Culture: A Reader, Second Edition.* New York: Routledge.

Covello, L. with G. D'Agostino 1958. *The Heart Is the Teacher: The Moving Story of an Immigrant Boy Who Became a Great Teacher.* New York: McGraw-Hill.

Crick, M. 1995. The Anthropologist as Tourist: An Identity in Question. In M.F. Lanfant, J. Allcock, E. Bruner (eds), *International Tourism: Identity and Change.* London: Sage Publications, pp. 205–23.

D'Agostino, F. and F. Vespasiano 2000. *L'icona della Sofferenza: Simbolismo del Corpo e Dinamiche di Gruppo nel Pellegrinaggio dei 'Battenti' alla Madonna dell'Arco.* Rome: Edizione Studium.

D'Aurea, D. 2010. Is Pasta Becoming the New Potato? The Rise in Popularity of Pasta in Ireland. In Patricia Lysaght (ed.), *Food and Meals at Cultural Crossroads.* Oslo: Novus Press, pp. 159–71.

Del Baldo, M. and P. Demartini 2012. Bottom-Up or Top-Down: Which Is the Best Approach to Improve CSR and Sustainability in Local Contexts? Reflections from Italian Experiences *Journal of Modern Accounting and Auditing,* 8(3).

Denzin, N.K. 1994. Chan is Missing: The Asian Eye Examines Cultural Studies. *Symbolic Interaction* 17: 63–89

Di Giovine, M. 2009a. *The Heritage-scape: UNESCO, World Heritage and Tourism.* Lanham: Lexington Books.

Di Giovine, M. 2009b. Revitalization and counter-revitalization: tourism, heritage, and the Lantern Festival as catalysts for regeneration in Hoi An, Viet Nam. *Journal of Policy Research in Tourism, Leisure and Events* 1(3): 208–30.

Di Giovine, M. 2010. La Vigilia Italo-Americana: Revitalizing the Italian-American Family through the Christmas Eve "Feast of the Seven Fishes." *Food and Foodways,* 18(4): 181–208.

Di Giovine, M. 2012. Padre Pio for Sale: Souvenirs, Relics, or Identity Markers? *International Journal of Tourism Anthropology* 2(2): 108–27.

Di Giovine, M. 2013. World Heritage Objectives and Outcomes. In Smith, C. (ed.). *Encyclopedia of Global Archaeology.* Vol. 11. NY: Springer, pp. 7894–903.

Di Giovine, M. 2014. The Imaginaire Dialectic and the Refashioning of Pietrelcina. In N. Salazar and N. Graburn. *Tourism Imaginaries: Anthropological Approaches*. Oxford: Berghahn, pp. 147–71.

Dietler, M. 2010. *Archaeologies of Colonialism: Consumption, Entanglement, and Violence in Ancient Mediterranean France*. Berkeley: University of California Press.

Eller, J.D. 1999. *From Culture to Ethnicity to Conflict: An Anthropological Perspective on Ethnic Conflict*. Ann Arbor: University of Michigan Press.

Eriksen, T.H. 1992. *Us and Them in Modern Societies: Ethnicity and Nationalism in Mauritius, Trinidad and Beyond*. Oslo: Scandinavian University Press.

Ferraresi, G. 2011. I volantini delle pizzerie da asporto a Milano e il menu di Padre Pio. *02blog.it*, November 11. http://www.02blog.it/post/9011/i-volantini-delle-pizzerie-da-asporto-a-milano-e-il-menu-di-padre-pio (Accessed on 15 January 2014.)

Gabaccia, D. 1998. *We Are What We Eat: Ethnic Foods and the Making of Americans*. Cambridge, MA: Harvard University Press.

Giddens, A. 1991. *Modernity and Self-Identity*. Cambridge: Polity Press.

Goldman, A. 1992. "I Yam What I Yam": Cooking, Culture and Colonialism. In S. Smith and J. Watson (eds), *De/Colonizing the Subject: The Politics of Gender in Women's Autobiography*. Minneapolis: University of Minnesota Press, pp. 169–95.

Goody, J. (1982). *Cuisine and Class: A Study in Comparative Sociology*. Cambridge: Cambridge University Press.

Graburn, N. 1983. The Anthropology of Tourism. *Annals of Tourism Research* 10(1): 9–33.

Hafstein, V. Tr. 2009. Intangible heritage as a list: From masterpieces to representation, in L. Smith and N. Akagawa (ed.), *Intangible Heritage*. London: Routledge, pp. 93–111

Hall, S. 1999. Whose Heritage? Unsettling "The Heritage," Reimaging the Post-nation. *Third Text* 13(49): 3–13.

Handler, R. 1988. *Nationalism and the Politics of Culture in Quebec*. Madison: University of Wisconsin Press.

Heldke, L. 2001. Let's Cook Thai. Recipes for Colonialism. In S. Inness, *Pilaf, Pozole and Pad Thai: American Women and Ethnic Food*. Amherst: University of Massachusetts Press.

Heldke, L. 2003. *Exotic Appetites: Ruminations of a Food Adventurer*. New York: Routledge.

Helstosky, C. 2004. *Garlic and Oil: Food and Politics in Italy*. Oxford: Berg.

Hewison, R. 1987. *The Heritage Industry: Britain in a Time of Decline*. London: Methuen.

Hobsbawm, E. and T. Ranger (eds) 1983. *The Invention of Tradition*. Cambridge: Cambridge University Press.

Johnston, J. and S. Baumann 2010. *Foodies: Democracy and Distinction in the Gourmet Foodscape*. New York: Routledge.

Junior League of Pasadena. 1976. *The California Heritage Cookbook*. New York: Doubleday.

Kirshenblatt-Gimblett, B. 1998. *Destination Culture: Tourism, Museums, and Heritage*. Berkeley: University of California Press.

Kirshenblatt-Gimblett, B. 2004a. Foreword. In L. Long (ed.), *Culinary Tourism: Material Worlds*. Lexington: University Press of Kentucky, pp. xi–xiv.

Kirshenblatt-Gimblett, B. 2004b. Intangible Heritage as Metacultural Production. *Museum International* 56, 1–2 (June 24): 52–65.

Levenstein, H. 1985. American Responses to Italian Food. *Food and Foodways* 1(1): 1–23.

Lévi-Strauss, C. 1963 *Structural Anthropology*. Trans. Claire Jacobson and Brooke Grundfest Schoepf. New York: Doubleday Anchor Books.

Long, L.M. (ed.) 2003. *Culinary Tourism*. Lexington: University Press of Kentucky.

Lowenthal, D. 1998. *The Heritage Crusade and the Spoils of History*. Cambridge: Cambridge

Lu, S. and G.A. Fine 1995. The Presentation of Ethnic Authenticity: Chinese Food as a Social Accomplishment. *The Sociological Quarterly* 36(3) (Summer): 535–53.

Maheu, R. 1966. *La Civilisation de l'universel*. Paris: Laffont-Gonthier

Matacena A. And M. Del Baldo (eds) 2009. *Responsabilità sociale d'impresa e territorio. L'esperienza delle piccole e medie imprese marchigiane*. Milan: F. Angeli.

Mauss, M. 2000. *The Gift: The Form and Reason for Exchange in Archaic Societies*. New York:

Mintz, S.W. 1986. *Sweetness and Power: The Place of Sugar in Modern History*. New York: Penguin Books.

Nagel, J. 1994. Constructing Ethnicity: Creating and Recreating Ethnic Identity and Culture. *Social Problems* 41(1): 152–76.

Norberg-Schultz, C. 1980. *Genius Loci: Towards a Phenomenology of Architecture*. New York: Rizzoli.

Olive Garden 2014. Culinary Institute of Tuscany: About Us. http://www. olivegarden.com/About-Us/Culinary-Institute-of-Tuscany/ (Accessed on 11 January 2014.)

Pagliai, V. 2009. Anti-racist Activism, Attempts towards Inclusive Citizenship, 'Foreigners Councils' and the Re-inclusion of the Race/National Divide in Tuscany. Address given at the Anthropology of Europe Workshop, University of Chicago, January 29, 2009.

Picard, D. and M. Di Giovine (2014). *Tourism and the Power of Otherness: Seductions of Difference*. Bristol: Channel View Publications.

Ram, U. 2007. Liquid Identities: Mecca Cola versus Coca-Cola. *European Journal of Cultural Studies* 10(4): 465–84.

Rigante, E. 1995. *Elodia Rigante's Italian Immigrant Cooking*. Cobb, CA: First View Books.

Ritzer, G. 2012. *The McDonaldization of Society: 20th Anniversary Edition*. New York: SAGE Publications.

Robertson, R. 1992. *Globalization: Social Theory and Global Culture.* London: Sage.

Sahlins, M. 2000. "Goodbye to Tristes Tropes." In *Culture and Practice: Selected Essays.* New York: Zone Books.

Salazar, N. 2013. Imagineering Otherness: Anthropological Legacies in Contemporary Tourism. *Anthropological Quarterly* 86(3): 669–96.

Salazar, N. and N. Graburn 2014. *Tourism Imaginaries: Anthropological Approaches.* Oxford: Berghahn Press.

Schlosser, E. 2012. *Fast Food Nation: The Dark Side of the All-American Meal.* Boston: Mariner Books.

Sia, M. 2013. *Mary Sia's Classic Chinese Cookbook.* Honolulu: University of Hawai'i Press.

Shortridge, B. 2004. Ethnic Heritage Food in Linsborg, Kansas and New Glarus, Wisconsin. In Long, L.M. (ed.), *Culinary Tourism.* Lexington: University Press of Kentucky, pp. 268–96.

Tornatore, J. 2012. Anthropology's Payback: "The Gastronomic Meal of the French." The Ethnographic Elements of a Heritage Distinction. In R. Bendix, A. Eggert and A. Peselmann (eds), *Heritage Regimes and the State.* Göttingen: Universitätsverlag Göttingen, pp. 341–65.

Trinh, M. 1991. *When the Moon Was Red: Representation, Gender and Cultural Politics.* London: Routledge.

Trubek, A.B. 2008. *The Taste of Place: A Cultural Journey Into Terrior.* Berkeley: University of California Press.

UNESCO 1945. *Constitution of the United Nations Educational, Scientific and Cultural Organization.* 16 November 1945. London.

UNESCO 1972. *Convention Concerning the Protection of the World Cultural and Natural Heritage.* 16 November 1945. Paris: UNESCO.

UNESCO 1989. Recommendation on the Safeguarding of Traditional Culture and Folklore. *Records of the General Conference*, Twenty-fifth Session Paris, 17 October to 16 November 1989. Appendix I, Part B., pp. 238–43.

UNESCO 1993. Establishment of a system of 'living cultural properties' (living human treasures) at UNESCO. (142 EX/18), UNESCO Executive Board meeting, 142nd session, Paris: UNESCO.

UNESCO 1994. Expert meeting on the 'Global Strategy' and thematic studies for a representative world heritage list. 20–22 June 1994. (WHC-94/CONF.003/INF.6). Paris: UNESCO.

UNESCO 1996. *Our Creative Diversity. Report of the World Commission on Culture and Development.* Paris: UNESCO.

UNESCO 2003a. Convention for the safeguarding of the intangible cultural heritage. 17 October 2003. Paris: UNESCO.

UNESCO 2003b. *UNESCO: What it is, what it does.* Brochure. Paris: UNESCO.

UNESCO 2004. The Yamato declaration for the integrated approach towards safeguarding tangible and intangible cultural heritage. (WHC-04/7 EXT.COM/INF.9). 7th Extraordinary Session of the 28th World Heritage Committee

Meeting, 6–11 December 2004. Paris: UNESCO UNESCO 2008. *World Heritage Information Kit.* Paris: World Heritage Centre.

UNESCO 2010a. Gastronomic Meal of the French. Decision 5.COM. 6.14. *Fifth Intergovernmental Committee Meeting for the Safeguarding of the Intangible Cultural Heritage.* Nairobi, Kenya, 15–19 November 2010.

UNESCO 2010b. Gingerbread Craft from Northern Croatia. Decision 5.COM. 6.10. *Fifth Intergovernmental Committee Meeting for the Safeguarding of the Intangible Cultural Heritage.* Nairobi, Kenya, 15–19 November 2010.

UNESCO 2010c. The Mediterranean Diet. Decision 5.COM. 6.41. *Fifth Intergovernmental Committee Meeting for the Safeguarding of the Intangible Cultural Heritage.* Nairobi, Kenya, 15–19 November 2010.

UNESCO 2010d. Traditional Mexican cuisine—authentic, ancestral, ongoing community culture, the Michoacan paradigm. Decision 5.COM. 6.30. *Fifth Intergovernmental Committee Meeting for the Safeguarding of the Intangible Cultural Heritage.* Nairobi, Kenya, 15–19 November 2010.

UNESCO 2013. Washoku, traditional dietary cultures of the Japanese, notably for the celebration of New Year. Decision 8.COM. 8.17. *Eighth Intergovernmental Committee Meeting for the Safeguarding of the Intangible Cultural Heritage.* Baku, Azerbaijan, 2–7 December 2010.

Van Esterik, P. 2006. From Hunger Food to Heritage Foods: Challenges to Food Localization in Lao PDR. In R. Wilk (ed.), *Fast Food/Slow Food: The Cultural Economy of the Global Food System.* Lanham: AltaMira Press, pp. 83–96.

Vázquez Montalbán, A. 1977. *L'art del menjar a Catalunya.* Barcelona: Edicions 62.

Watson, W. 2006. *Golden Arches East: McDonalds in East Asia.* 2nd Edition. Palo Alto: Stanford University Press.

Wilk, R. (ed.) 2006. *Fast Food/Slow Food: The Cultural Economy of the Global Food System.* Lanham: AltaMira Press.

Wilk, R. 1999. "Real Belizean Food": Building Local Identity in the Transnational Caribbean. *American Anthropologist* 101: 244–55.

Chapter 1

Re-Inventing a Tradition of Invention: Entrepreneurialism as Heritage in American Artisan Cheesemaking

Heather Paxson

In his 2006 book, *The United States of Arugula*, David Kamp credits Laura Chenel with almost singlehandedly introducing goat cheese to America by becoming its first domestic commercial producer (2006: 171–2). Chenel's story, in his telling, contains two iconic features. First, to perfect her craft, Chenel traveled to France, the quintessentially cultured epicenter of "real" cheese. Second, she got her commercial break in 1980 when she drove a batch of fresh cheeses from her Sonoma County goat farm to Berkeley and walked into Chez Panisse restaurant. Alice Waters put the chèvre in a salad, named Chenel on the menu, and the rest is … invented tradition? Today the story of Laura Chenel's chèvre is often told to establish the beginning of a current "renaissance" in American artisan cheesemaking. A handful of Americans involved in the back-to-the-land movement, including Chenel, began making cheese by hand for commercial sale in the early 1980s; since 2000, the number of domestic artisan producers has more than doubled. The heroine of the chèvre's story embodies characteristics that could describe the American artisan industry as a whole: it is innovative, it is entrepreneurial and it borrows unapologetically from European tastes and savoir-faire.

Drawing on several years of ethnographic research among artisan cheesemakers, retailers and boosters in the United States, I suggest that a tradition of artisanal cheesemaking—a tradition that is consciously cultivated as newly emergent—locates its distinctively American heritage neither in the taste of a food nor in customary practice of cheese fabrication, but instead in an entrepreneurial sensibility. At talks given at annual meetings of the American Cheese Society and in a raft of recent popular cheese books, American artisanal cheesemaking is collectively characterized by shared cultural commitment to *innovation* as a source of value and integrity—even artisanal authenticity. What makes an American cheese distinctively American (in this formulation) is that it presents itself as new, different, unique—despite the fact that, as acknowledged in Laura Chenel's story of pilgrimage to France, it remains inescapably indebted to European histories of practice and taste-making. Often obscured in this formulation is how today's innovators are also indebted to American histories of practice and taste-making that have unfolded in New England, Wisconsin, central California, and elsewhere across the States (Paxson 2010).

To frame my argument about American invented traditions of artisanal cheesemaking innovation, I will first briefly characterize the foil against which claims to American exceptionalism are made: namely, invented traditions of European continuity. Just as the appearance of continuity in European food traditions relies on changing methods of fabrication and marketing, I will demonstrate through the example of the oldest continuously operating artisan cheese factory in the United States how an emphasis on change relies on continuity "in order to demonstrate its effect" (Strathern 1992: 3).

French Cheese and the Invention of Tradition

In the course of my research among American cheesemakers, I lost count of how many times I heard repeated General Charles De Gaulle's complaint, "How can you govern a country which has 246 varieties of cheese?" (Mignon 1962) (The number of cheeses changed with each recitation.) De Gaulle's quip tells us something about the role of food, and of cheese in particular, in helping to establish a collective sense of regional affiliation. Cheese, an ancient, domestic means of preserving that most perishable agrarian product, milk, would seem ready-made to embody cultural heritage. De Gaulle's 246 (or so) varieties of cheese work as a symbol of national unruliness only if we imagine each of those cheeses as emerging from a politically entrenched patchwork of customarily distinct regions. De Gaulle's task was to unify a people loyally committed not to the excellence of "French" cheese but to 246 regionally distinctive cheeses and the cultural heritage for which they stand.

One of France's most elaborate and successful cheese traditions concerns Camembert, said to have originated when a Norman farmwoman named Marie Harel followed a "secret" recipe for Brie using a smaller Livarot mold (practicing the sort of improvisational tinkering I have found among American cheesemakers) and trained her children and grandchildren to carry on making the cheese as family patrimony. As Pierre Boisard (2003) details in *Camembert: A National Myth*, the cheese's story begins to transcend Norman regionalism and to take on the significance of a national myth because the tale is set in the early years of the French Revolution (1791) and because the secret Brie recipe is said to have been given to Madame Harel by a priest (who stands in the story as a representative of the *Ancien Régime*) seeking refuge with the Harel family while fleeing persecution by the revolutionaries. Thanks to Madame Harel's entrepreneurial industry, writes Boisard, "a bit of old France, of pre-Revolutionary France, will survive" into the future in a new form (2003: 10).

The myth of Camembert fits neatly Eric Hobsbawm's formula for "invented traditions," which he describes as "responses to novel situations which take the form of reference to old situations, or which establish their own past by quasi-obligatory repetition" (1983: 2). No surprise, then, to learn that Marie Harel's mythological fame does not, in fact, date back to the Revolution. Instead, 130 years went by before the American Joseph Knirim turned up in the town of Camembert (pop. 300) to venerate the memory of Madame Harel and her "veritable Norman Camembert" by erecting a statue in her honor. Knirim, a physician, adulated Marie Harel's cheese not for its taste and sumptuousness, but for its "digestibility." In a letter to the townspeople of Camembert, Knirim explained: "Years ago, I suffered for several months from indigestion, and Camembert was practically the sole nourishment that my stomach and intestines were able to tolerate. Since then, I have sung the praises of Camembert, I have introduced it to thousands of gourmets, and I myself eat it two or three times a day" (Boisard 2003: 3). Only once the visiting American had erected her statue did Marie Harel's name begin to stand for the essential contribution that peasant agriculture has made to the strength of the French nation. Camembert's iconic Frenchness seems little tarnished by having become one of the nation's most industrialized cheeses. Long seeded with laboratory-isolated strains of *Penicillium candidum* to produce a pure-white coat of mold, and now most often made from pasteurized milk, Camembert's materiality today is only a shadow of what once cured Knirim's indigestion. In hopes of recuperating at least a hint of that past, Norman dairy farmers and cheesemakers have secured *Appellation d'Origine Contrôlée* (A.O.C.) status not for Camembert per se, but for Camembert de Normandie. For a cheese to qualify for the site-specific name, production must occur within geographically limited areas and comply with a voluminous set of regulatory standards (Rogers 2008). Norman Camembert and French Camembert now vie for consumer sentiment and market position. Throughout Europe, what heritage food will look like in the future is a contentious matter of politics and policy to be worked out through the legal instruments of geographical indications (on cheese, see: Boisard 1991, 2003, Grasseni 2003, 2009, Rogers 2008, on wine, see Ulin 2002, Guy 2003, Demossier 2011).

Such well-cultured cheeses as Camembert, Comté, and Taleggio are usefully analyzed as embodying and reproducing the "invented traditions" of country idylls populated by an immemorial peasantry (Boisard 2003, Rogers 2008, Trubek and Bowen 2008, Grasseni 2009). When a new wave of American cheesemakers travels to France to learn how to make "real" cheese, they reinforce European inventions of culinary tradition as authentic and gastronomically superior. Without doubt, many excellent cheeses are made in France and throughout Europe—but so, too, are boring supermarket cheeses. It is a testament to the success of France's invented cheese traditions and to the branding of French cheeses as fundamentally authentic and traditional—even when most Camembert today is, in fact, made from pasteurized milk and ladled by robots—that foil-wrapped, processed wedges of Laughing Cow (*La Vache Qui'rit*, in its native tongue) are never metonymically dubbed "French cheese" (Boisard 1991). As Hobsbawm writes, "It is the contrast between the constant change and innovation of the modern world and the attempt to structure at least some parts of social life within it as unchanging and invariant, that makes the

'invention of tradition' so interesting for historians of the past two centuries"
(1983: 2). In Europe, where "the traditional" and "the modern" continue to be
potent, mutually constitutive tropes through which people stake moral claims of
belonging, authenticity, and progress, "invention of tradition" is a particularly
useful analytic (cf. Terrio 2000). But in the United States, where progress is
valued over patrimony, what is invented as tradition—what is enshrined as a
matter of cultural heritage—is continual change and innovation, not continuity.
In the United States, continuity in practice, in know-how, in form, risks being
labeled old-fashioned or, worse, boring, and so continuity is often obscured in
narratives of innovation. Americans, ever impatient for a brighter tomorrow, are
continuously remaking and marketing their traditions as new, fresh, and exciting.

American Cheese and a (Continuously Re-invented) Tradition of Invention

In 1865, with Lincoln in the White House and the Civil War just coming to an
end, the Marin French Cheese Company began making cheese (originally the
Thompson Brothers Cheese Co.) after Jefferson Thompson, a dairy farmer,
recognized an emergent market niche in the port town of San Francisco. Compare
to the above story of Camembert that of Marin French, the oldest continuously
operating cheese factory in the United States, located in Petaluma, California. In
a 2008 interview, the late Jim Boyce, who in 1998 purchased the company from
Thompson's descendents, told me the story as he learned it from an employee who
had just retired after 60 years with the company.

During the California Gold Rush (1849–1855), the story goes, European
stevedores (deckhands) who sailed into Yerba Buena harbor (later, San Francisco
Bay) delivering goods to support the mining enterprises got "caught up in the
fever" and abandoned ship to seek their own fortune in the mines. After the gold
rush went bust, workers returned to the bay to make a living at the dockyards.
Boyce said to me:

> Now, in any workman's bar or inn … you work hard, you get dehydrated, you
> go to the bar for hydration and energy—most typically that's given to you by
> beer so you can quickly restabilize yourself. … The beer gives them hydration
> and carbohydrate but no protein. And most typically in a workman's bar
> there's a jar of pickled eggs or something like that, pig knuckles, sausage. [But
> here] there weren't any eggs; no chickens— nothing had been developed. …
> Well, Jefferson Thompson, the dairyman on this farm [the site of the present-
> day factory] says to himself in a moment of marketing brilliance, 'I wonder
> if they'd eat cheese, instead?' So he starts making these little cheeses, three-
> ounce cheeses, more or less. And he hauls them off to the docks, and they put
> them on the table in a bowl, and they were an immediate hit! Why? Because
> these are European stevedores: they knew cheese. They ate it breakfast, lunch,
> and dinner. And that was the origin of the company.

Whereas Europe's invented food traditions mean to legitimate present practices by claiming continuity with the past (even if shaped in a new form), American origin stories mark decisive breaks with the past. The story of Laura Chenel chèvre exemplifies individual passion and entrepreneurial opportunism; Marin French's Breakfast Cheese celebrates the creation of new markets.

A tradition of invention is enshrined in the American Cheese Society (ACS) designation of American Originals as a classificatory category for its annual judging and competition. American Originals designate cheeses invented on American soil: Colby and Brick, invented by first-generation immigrants in nineteenth-century Wisconsin; Teleme and Jack, invented in twentieth-century California. In recent years, the American Cheese Society has added "original recipe" subcategories of American Originals; 2011 award winners included Mt. Tam, Cocoa Cardona and Flagsheep. The theme of the 22nd Annual Meeting of the ACS held in Louisville, Kentucky, in 2005—*Creating Tradition*—did not so much offer a self-conscious look at how American cheese traditions had been invented as it set out to create tradition anew from this point onward, into the future. The call for Americans to create a cheesemaking tradition arises from a feeling among newer artisans, those getting their start since Laura Chenel's goat cheese hit Berkeley, that they have been largely on their own, starting from scratch.

Allison Hooper, co-founder of the 25-year-old Vermont Butter & Cheese Company, writes in her Foreword to Roberts' *Atlas of American Artisan Cheese*, "Without the burden of tradition we are free to be innovative, take risks," suggesting that a lack of tradition in regional cheese types and fabrication method is a virtue rather than a deficit because it opens up possibilities for experimentation (2007: xiii). Dancing Cow's Sarabande, a cow's milk cheese with a washed rind, is molded in a truncated pyramid form, the kind used in France for Valençay, a charcoal-dusted goat's milk cheese from the province of Berry (legend has it that the cheese was once a made in a perfect pyramid until Napoleon, passing through Valençay town following a failed military campaign in Egypt, was so enraged by the cheese's taunting shape that he lopped off the top with his sword, leaving the form that survives today). In a presentation at the 2007 American Cheese Society meetings, Steve Getz, co-owner and operator of Dancing Cow (which has since gone out of business), delighted in announcing that it had been recently declared illegal in France to make a cow's milk cheese in a truncated pyramid form (the shape is reserved for goat's milk cheese).[1] At an ACS panel the following year devoted to "European Forebears: Reinventing the Classics," Flavio DeCastilhos, who started a farmstead cheese operation after leaving a successful Internet start-up, described the reaction of a Dutch cheesemaking consultant he brought to Oregon to develop a line of Gouda-style cheeses for his Tumalo Farms:

---

1   Chapter 5, Article 2 of France's "Decree no. 2007-628 relating to cheeses and specialty cheese houses" (April 27, 2007) stipulates that "pyramid and truncated pyramid" forms are "reserved exclusively for goat's milk cheeses."

> I had this really interesting idea that I wanted to make this cheese— I want to
> have this hoppy flavor, I want to put beer in it. So [he] turns to me and says, 'I
> can't help you.' I said, 'Why not?'
> 'Well, in Holland, we drink the darn beer. You're on your own.'
> So I had to go and figure it out myself. But that's how the Pondhopper was born.[2]

Cheesemakers develop original product lines by tinkering with established
recipes and bestowing novel names on resulting cheeses: use goat's milk rather
than cow's in a Gouda recipe, wash it with a local microbrew beer, and call the
cheese Pondhopper; or, start with a Havarti recipe but blend sheep and cow's
milk to come up with Timberdoodle (on naming cheeses, see West et al. 2012).
Such cheeses exemplify how American artisans, unconstrained by expectations
for fidelity to customary form, seek to redefine "American cheese" by creating a
tradition of invention.

Despite persistent claims to novelty, my point is that this sentiment and practice
has a history. In our interview, Jim Boyce explicitly likened the present era to that
of turn of the twentieth century in terms of patterns of cheese consumption as well
as artisan modes of production. The early 1890s, he explained, saw a flourishing of
cheesemaking activity not only in New England and the upper Midwest, but also in
port cities up and down the Pacific coast; among the more successful was Tillamook,
which first opened in Oregon in 1894. A similar flourishing of cheesemaking
enterprises began again in the 1990s: two years after its founding in 2006 the Oregon
Cheese Guild boasted 13 members, including Tillamook. Both eras, Boyce said,
have been periods "of innovation in local cheese."

Marin French, Boyce told me, "survived wars, it survived depressions, it
survived dot coms, it survived what I call the Cheese Depression, which was in
the early eighties—it was discovered that cheese had fat and … if you had fat in
your food, it was no good." Marin French's success has been made possible by
the strong cheese market of San Francisco, but that market has had to be marketed
to. Innovation is not to be romantically imagined as a craftsperson's singular
artistic creativity. Successful entrepreneurial innovation responds to customer
tastes that transform alongside demographic shifts (immigration, urbanization,
class mobility) and broader culinary trends. As Howard Becker points out, craft's
defining utility implies that its objects and activities must be useful to *others:*
"If a person defines his work as done to meet someone else's practical needs,
then function, defined externally to the intrinsic character of the work, is an
important ideological and aesthetic consideration" (1978: 864). Consumer desire
helps to constitute craft not only by providing necessary markets, but also through
informing aesthetic standards.

Having got its start selling small rounds of "Breakfast Cheese" to European
deckhands to accompany their morning ale, by the turn of the twentieth century

---

2 25th Annual meeting of the American Cheese Society, Chicago, Illinois,
July 25, 2008.

Marin French had introduced an Austrian-style, smear-ripened cheese called Schloss. In the early 1900s they launched Thompson Brothers Camembert: hand-molded Camembert was produced in Marin County prior to the Great San Francisco earthquake. In 1907, one year after the earthquake, Thompson renamed the Camembert "Yellow Buck." Boyce speculated that Yellow Buck was named after the buck elk that were at one time plentiful in the area and that are now being reintroduced in a sanctuary on Point Reyes peninsula. At the same time, Yellow Buck is "a symbol of strength" that Boyce interpreted as speaking to the rugged beauty of the Marin landscape, as well as being "sort of masculine." A company that got its start selling cheese in saloons seems consciously to have worked to sustain a masculine image in marketing its cheese with male as well as female consumers in mind. Masculine appeal, like any other culturally meaningful symbolic marker, is not a static quality. The Yellow Buck label was retired in the teens or twenties and replaced by the regal, Frenchified brand name, Rouge et Noir. The entrepreneurial spirit in evidence in today's artisan "renaissance" is part of and indebted to a longstanding tradition of innovation. I have argued, moreover, that the American pioneering ideal has contributed to a collective neglect of an ongoing history of artisan cheesemaking, one long characterized by innovation in marketing as well as craft method (Paxson 2010).

Getting product to market at a viable price is essential to any commercial enterprise. Describing to me how the Thompson brothers once transported fresh cheeses by horse and wagon to the Petaluma River and then by steamer across the Bay, Jim Boyce offered this analysis:

> ... it's putting a product together with a very receptive group of people who understood and could enjoy the product. It's pure marketing— it's marketing at its greatest! It's the individual who's saying, 'What if?' ... I also think that it is part of the foundation of why today San Francisco is the strongest cheese market in the country. I think you can take its roots right back to the day cheese was delivered to the docks of San Francisco, to the workers.

By locating the authenticity of a food in its history of "pure marketing," Boyce offered a savvy cultural analysis. In her Master's thesis, "Chore, Craft and Business: Cheesemaking in 18th Century Massachusetts," Kristina Nies describes how eighteenth-century cheesemakers "made adjustments for seasonal fluctuations as well as for the marketplace" (2008: 10). She writes of one Massachusetts farmwoman, Elizabeth Porter Phelps, who innovated a recipe for full-fat cheese after her husband, who marketed the cheese in Boston, reported that it would command a higher price than the usual skimmed-milk cheese (Nies 2008: 10). Deborah Valenze writes similarly of eighteenth-century English dairywomen, "Long accustomed to selling their products, if only on a local basis, they showed considerable sensitivity to the ever-elusive predilections of the market" (1991: 154). In the early 1800s, Phelps, like Thompson a hundred years later—and like Swiss immigrant John Jossi, who developed Brick cheese for German settlers in 1870s Wisconsin—was a cheese innovator driven by commercial possibility.

Jim Boyce did not innovate in the vat; his entrepreneurial creativity was expressed through marketing. Boyce, who reintroduced the Yellow Buck label in 2000 to commemorate [roughly] the Camembert's 100-year anniversary, took over the company amidst a financial slump. To turn things around, not only did he diversify the product line by introducing blue-veined and flavored varieties as well as goat's milk cheeses, he revised the company's marketing strategy. When he acquired the company, Marin French was producing Brie and Camembert in small batches from hand-cut curd, manually bucketed, aged, and individually wrapped by hand—and it was all sold in the deli dairy. As Jim Boyce schooled me, commercial dairy products are sold in supermarkets as either deli dairy or service deli. Deli dairy refers to the large refrigerated cases along the back walls of supermarkets that contain perishable staples: butter, milk, yogurt, and blocks of "everyday" cheese, encased in plastic. Today, while Marin French still owns supermarket shelf space in regional stores, most of its product is now directed to the service dairy and displayed as a specialty item in a center island, perhaps, rather than as a staple food alongside milk and butter. Boyce began to market the cheese in a way that calls attention to its method of fabrication, bestowing the cultural cache of "artisanal" on a label that has been around for over a hundred years but must now be "discovered" by an emergent, discriminating cheese world. Cristina Grasseni writes of Italian cheesemakers trying to capitalize on consumer interest in heritage foods, "the commercialization of cheese as a traditional product entail[s] not only a transformation of traditional skills but also the acquisition of new skills for managing one's image" (2003: 260; see also Terrio 1999). What Grasseni calls "packaging skills" are, in Italy, hidden behind appeal to heritage, but in the U.S., they are as likely to be announced as evidence of authentic entrepreneurial acumen.

Conclusion

In suggesting that in the United States artisanal cheese is better characterized in terms of a "tradition of invention" (albeit invented as such) in contrast to the "invention of tradition," I have meant to point out the cultural influence of the American celebration of entrepreneurial innovation on craft industry and food heritage. While Europe's traditions may not be as old as some imagine, so too may America's inventions not be so innovative. American cheesemakers may not feel burdened by the bureaucratic constraints of Europe's government-protected "traditional" recipes, but when they dream up new cheeses they inevitably modify old ones. As Marquis and Haskell write in *The Cheese Book*, "the story of cheesemaking is a long history of imitation, ever since the first cheeses were made" (1964: 18). A Wisconsin cheesemaker who recently started adding value to his German-American wife's family's 160-year-old dairy farm by making Italian-style cheeses, acknowledged to me in an interview that "the technical part of making cheese ... we're grabbing, stealing, borrowing" from European models. Tumalo Farms' Pondhopper, on this view, is not so much an American Original as fancy Gouda. When newness is imagined as the source of a good's value, people can end up reinventing wheels of cheese.

If we view food heritage as a set of socially reproduced standards of taste as well as practical know-how, it is quite possible to trace continuity as well as change in American cheesemaking. "American" taste has always reflected the backgrounds of immigrant communities who have brought with them culinary preferences as well as dairying and cheesemaking skills (not to mention livestock), more recently from Latin America and previously from Europe: the Dutch in New York; Germans in Pennsylvania and later Wisconsin; English in New England. Such regional variation, Nies observes (2008: 35), might help explain the disputes over proper cheesemaking procedure evidenced in texts published in the 1790s in which authors disagree over the propriety of coloring cheddar-style cheese orange with the annatto seed. Still today, orange cheddar remains standard in Wisconsin and disdained in New England and New York, suggesting a kind of regional place-based heritage for at least this English-American variety of cheese. When the new artisan entrepreneurs wanting to make a mark with their singular creations overlook this *as* cheese heritage, they do so to establish their own artisanal authenticity as innovative and thus properly American.

References

Becker, H.S. 1978. Arts and Crafts. *American Journal of Sociology* 83(4), 862–89.

Boisard, P. 1991. The Future of A Tradition: Two Ways of Making Camembert, the Foremost Cheese of France. *Food and Foodways* 4(3–4), 173–207.

Boisard, P. 2003. *Camembert: A National Myth*. Translated by Richard Miller. Berkeley: University of California Press.

Demossier, M. 2011. Beyond *Terroir*: Territorial Construction, Hegemonic Discourses, and French Wine Culture. *Journal of the Royal Anthropological Institute* 17, 685–705.

Grasseni, C. 2003. Packaging Skills: Calibrating Cheese to the Global Market, in *Commodifying Everything: Relationships of the Market*, edited by Susan Strasser. New York: Routledge, 259–88.

Grasseni, C. 2009. *Developing Skill, Developing Vision: Practices of Locality at the Foot of the Alps*. Oxford: Berghahn Books.

Guy, K.M. 2003. *When Champagne Became French: Wine and the Making of a National Identity*. Baltimore: Johns Hopkins University Press.

Hobsbawm, E. 1983. Introduction: Inventing Traditions, in *The Invention of Tradition*, edited by E. Hobsbawm and T. Ranger. Cambridge: Cambridge University Press, 1–14.

Hooper, A. 2005. The Business of Farmstead Cheesemaking, in *American Farmstead Cheese: The Complete Guide to Making and Selling Artisan Cheeses*, edited by Paul Kindstedt. White River Junction, VT: Chelsea Green, 227–46.

Kamp, D. 2006. *The United States of Arugula: The Sun-Dried, Cold-Pressed, Dark-Roasted, Extra Virgin Story of the American Food Revolution*. New York: Broadway Books.

Marquis, V. and P. Haskell. 1964. *The Cheese Book: A Definitive Guide to the Cheeses of the World*. New York: Simon & Schuster.

Mignon, E. 1962. *Les Mots du Général*. Paris: A. Fayard.

Nies, K. 2008. Chore, Craft and Business: Cheesemaking in 18th Century Massachusetts. M.A. Thesis, Boston University.

Paxson, H. 2010. Cheese Cultures: Transforming American Tastes and Traditions. *Gastronomica: The Journal of Food and Culture* 19(4), 442–55.

Roberts, J P. 2007. *The Atlas of American Artisan Cheese*. White River Junction, VT: Chelsea Green.

Rogers, J. 2008. The Political Lives of Dairy Cows: Modernity, Tradition, and Professional Identity in the Norman Cheese Industry, Ph.D. Dissertation, Brown University.

Strathern, M. 1992. *After Nature: English Kinship in the Late Twentieth Century*. Cambridge: Cambridge University Press.

Terrio, S.J. 1999. Performing Craft for Heritage Tourists in Southwest France. *City and Society* 11(1–2), 125–44.

Terrio, S.J. 2000. *Crafting the Culture and History of French Chocolate*. Berkeley: University of California Press.

Trubek, A.B. and S. Bowen. 2008. Creating the Taste of Place in the United States: Can We Learn from the French? *GeoJournal* 73, 23–30.

Ulin, R. 2002. Work as Cultural Production: Labour and Self-Identity among Southwest French Wine Growers. *The Journal of the Royal Anthropological Institute* 8(4), 691–712.

Valenze, D. 1991. The Art of Women and the Business of Men: Women's Work and the Dairy Industry. *Past and Present* 130(1), 142–69.

West, H.G., H. Paxson, J. Williams, C. Grasseni, E. Petridou and S. Cleary. 2012. Naming Cheese. *Food, Culture and Society* 15(1), 7–41.

Chapter 2

Terroir in D.C.? Inventing Food Traditions for the Nation's Capital

Warren Belasco

When, in the exuberant early days of the Obama presidency, Michelle Obama planted her vegetable garden on the White House lawn and sponsored a new farmer's market nearby, it seemed that the local food movement had come to official Washington. Even before his inauguration the President-Elect had made a pilgrimage to Ben's Chili Bowl, a down-home hot dog stand known mainly to its indigenous black residents, and this too prompted speculation that the Federal City's food resources might finally be receiving some respect. Might D.C., a city historically oblivious, if not overtly hostile to its own roots, culture, and environment, have its own *terroir*?

The question was not frivolous, for as Amy Trubek argues in *The Taste of Place* (2008), the possibilities for and implications of *terroir* extend far beyond its origins as an assertion of French culinary superiority—a hegemonic claim familiar to status-conscious Americans, especially in Francophile official Washington (Landau 2007, Haley 2011, Strauss 2011). But when broadened to mean a pride in place—*any* place—*terroir* takes on a more functional meaning that transcends the invidious distinctions of the culinary authenticity game. Rather than establishing hierarchies of taste, the newer, democratized *terroir* fosters the three Rs: *regard* for one's native landscape, *reciprocity* between food producers and consumers, and an overall sense of *responsibility* for the consequences of one's own behavior. In short, as the theory goes, pride in place becomes an instrument of local environmental, economic, and cultural regeneration (Russo 2009). And if this could happen in the nation's capital, long the test case for social reform, it would bolster the local foods trend everywhere.

Establishing a sense of *terroir* in the nation's political capital was no easy feat, however. There is a difference between living in a place and loving that place. Even the more democratized, functional *terroir* needs a positive reputation as a base. Unlike other American culinary powerhouses such as New Orleans, New York, Chicago, and San Francisco, however, Washington did not have a wealth of native pride to draw upon. On the contrary, the Federal City was established in the 1790s to be distinctly national, modern, and placeless, a neutral site to which jealous, competing states could send representatives without fear of being overwhelmed by local interests and diversions (Young 1966). Carved out of the declining tobacco and wheat farms of Maryland and Virginia, and plagued from the start by land speculation, cost-overruns, and political neglect, D.C. had few indigenous boosters or traditions upon which to build (Luria 2006). Early travelers and residents alike were particularly scathing about the food, which was "atrocious .... not even any fruit fit for hogs" (Young 1966: 49). Years later, Horace Greeley (1868) repeated the still-familiar litany: "The rents are high, the food is bad, the dust is disgusting, and the morals are deplorable" (cited in Green 1962: 312). Although the cosmopolitan city that emerged in the late nineteenth century entertained millions of visitors yearly, it fed them largely in bland, chain-operated museum cafeterias of the kind that also sustained government workers. Not much *terroir* there! Even when Washington's booming restaurant scene began to receive national notice in the 1990s, the notice was grudging and came coupled with references to its mediocre culinary past. "For years," the *New York Times* carped in 1990, "well-traveled food mavens from New York and other gastronomic centers considered Washington about as provocative as a tax audit" (Miller 1990). The inferiority complex was often internalized by local reviewers, who framed their praise for current achievements with a tone of astonishment, as in: "If you haven't eaten in Washington recently, you're in for a monumental surprise," (Sietsema n.d.) and "D.C. area's cool, new reputation celebrated ..." (Belgacafe. com 2005). And even in newly "cool" D.C., most of the "hot" restaurants were of the transnational fusion variety found in most other cosmopolitan cities; native Washingtonians were hard-pressed to come up with dishes indigenous to the city, except perhaps for the "half-smoke" pork/beef sausage popularized by Ben's Chili Bowl. That Ben Ali, the Muslim proprietor, could not eat his own product was an apt symbol of his city's own self-alienation. Indeed, for more than two centuries, Washington has had to contend with a deeply-engrained tradition of disrespect that must be acknowledged and explained before we can attempt to construct a local food heritage supportive of the three Rs.

The Tradition of Disrespect

From a culinary standpoint, D.C. certainly got off on the wrong foot. Arriving from more settled places, new residents and visitors routinely bemoaned the limited amenities of this rudimentary "city of magnificent intentions" (Dickens 1842: 281). Although early presidents served decent food to those deemed worthy of an invitation, for daily provisioning Washington was a food desert dependent on crowded boardinghouses and rough taverns (Carson 1990). Crude accommodations and challenging summer weather induced politicians, staffers, and lobbyists to come to D.C. just for the cool winter season. Few brought wives, family, and personal cooks. This male-dominated seasonality established an enduring white transient "sojourner culture" indifferent to local roots (Allgor 2000, Young 1966, Jacob 1994). "Everybody is a bird of passage," an 1820s observer noted, "from the president down, and no one thinks of being at home there ..." (Bryan 1916: 62). Even longtime government workers were, and to a great extent, still are more like "permanent temporaries" with little attachment to D.C.'s history, culture, and landscape (Green 1962: II: viii). What guidebook author John Ellis wrote of federal clerks in 1869 could well apply to many of the region's suburbanized residents today: "They are strangers to the city ... and never become fully domesticated. They are in Washington, but not of it" (369). It should be noted that some black residents may have felt quite differently, as D.C. offered unparalleled economic opportunities for slaves and freedmen alike (Brown 1972, Provine 1973, Gatewood 2000, Anacostia Museum 2005), but their enthusiasm for the place hardly counted in official culture, and their culinary contributions were often dismissed, especially by Northerners, as when Ohio Senator Benjamin Wade observed in 1851 that Washington food "is all cooked by niggers, until I can smell and taste the nigger" (Genovese 1976: 540).

Northerners who would have preferred whiter, more "civilized" Philadelphia or New York to be the national capital viewed D.C. as a southern swamp, an image that was especially devastating—and somewhat inaccurate, as D.C.'s wetlands were more like grassy marshes than treed swamps (Miller 1989). But the latter captured the place's early desolation, starting with Rep. Ebenezer Mattoon in 1801: "This swamp—this lonesome dreary swamp, secluded from every delightful or pleasing thing ..." (Crew 1892: 107). Whereas Native Americans, subsistence farmers, and enterprising black and Irish workers foraged profitably in the area's nutrient-rich wetlands (Rice 2009), genteel observers saw a dangerous "miasmal" wasteland—a "fever-stricken morass," according to Henry Adams (1889: 30)—that enfeebled year-round residents, swallowed up wayward visitors, and corrupted politicians. The disjunction between the early city's neo-classical aspirations and its rougher realities was particularly galling to one 1804 observer:

This fam'd metropolis
Where fancy sees
Squares in morasses,
Obelisks in trees.
—Thomas Moore, 1804 (Green 1962: 39)

Even in 1983 Smithsonian American History Museum director Roger Kennedy referred to Washington's "grandeur in a swamp" (xviii). While some locales could make a gastronomic virtue of their wetland resources—Louisiana's bayous, New York's Sound, Minnesota's lakes, Baltimore's Chesapeake—D.C.'s swamps were more associated with the despised Irish and African American workers who built and maintained the city. Thus the Irish neighborhood was called "Swampoodle," and the Potomac marshes sheltered runaway slaves. Moreover as the river often flooded the official downtown and impeded development of massive offices and museums, its shores were drained, walled, and filled at first opportunity (Tilp 1978, Lessoff 1994). Such reclamation expanded D.C.'s monumental core at the expense of food-rich ecosystems. Conversely, as I will argue, reinventing local food traditions may entail restoring the "swamps," and the river that feeds them.

The determination to reclaim Washington from its wastelands and fulfill Pierre L'Enfant's original (1791) imperial design also led to downtown development projects that, in retrospect, robbed the city of culinary touchstones and traditions that have helped famous food cities like New Orleans and San Francisco establish and commodify their local food identities (Goode 2003). Here I will briefly discuss how the city's role as abstract exemplar of modern nationalism (Bednar 2006, Savage 2009) resulted in the loss of three food-rich nineteenth-cenury sites that, had they survived, might have served the heritage-hunger of post-modern generations.

Center Market: Perhaps reflecting L'Enfant's French roots, the original plan for D.C. set aside valuable acreage for the establishment of a central food market south of Pennsylvania Avenue between 7th and 9th Streets NW. Conveniently located next to a canal that connected downtown with the Potomac and Anacostia Rivers, the market attracted farmers from throughout the region and quickly became a major social hub, as residents of all classes and races—from Presidents to slaves—gathered regularly to shop and talk. One mid nineteenth-cenury handbook boasted,

> A greater variety of good things can nowhere be found collected under one roof, than may at all times be found in Center market. The highlands of Maryland and Virginia supply it with beef and mutton, that cannot be excelled, while the adjoining country pours into it a variety of vegetables that makes one wonder where they all come from. In the way of fish, the Potomac yields a great variety, the shad, rock fish or basse and the oysters, having no superior in the country; and no market is better supplied with the venison, wild turkey, ortolon, reed-birds and the famous canvas back ducks. (*Bohn's Hand-Book of Washington*, 1856)

Opened in 1871, Adolf Cluss' new Center Market building was one of the world's largest and most architecturally distinguished public markets—a turreted red brick palace with over 600 permanent stalls inside plus several hundred more for farmers outside. A hundred years later some cities would turn much smaller and plainer public markets into cornerstones of downtown redevelopment—one thinks of Seattle's Pike Place, Baltimore's Lexington Market, Boston's Quincy Market, Philadelphia's Reading Market, New Orleans' French Market, San Francisco's Ferry Building, and so on. But in the 1930s D.C.'s Center Market—losing trade to suburban groceries—was demolished to make way for a new National Archives, a mausoleum for noble documents—an almost too perfect example of the dualistic priority accorded mind over body (Tangires 1999, 2008).

Federal Triangle: A similar fate awaited the rest of the vibrant neighborhood located south of Pennsylvania Avenue between White House and Capitol. Here stood hundreds of taverns, boarding houses, rickety hotels, "gambling hells," cheap theatres, oyster houses, and, most suggestive, the red light district known after the Civil War as "Hooker's Division," a semi-permanent encampment of prostitutes originally recruited by General Joseph Hooker to "service" his Union Army troops (Ellis 1869, Bryan 1916). Like Center Market, all of this was demolished in the 1930s to make way for the colossal neo-classical government offices known as Federal Triangle. While it is impossible to assess the quality of food served in the lost area, some anecdotal and archaeological data does suggest that D.C.'s "Hooker's Division" did offer some of the best food in town (Seifert 1991, National Museum of the American Indian 1997, Evelyn & Dickson 1999). Indeed, a bordello theory of haute cuisine might argue that the "great" culinary reputations of some cities—San Francisco, New Orleans, more recently Las Vegas—were established by the highly competitive nature of the "sin" trade, in which proprietors hired top chefs to lure customers to their establishments. In Washington, however, all of this gave way to antiseptic Federal Triangle, an area dependent on government cafeterias by day and still largely vacant by night. Thanks largely to such wholesale sterilization, Washington's downtown remains, as described by the Smithsonian's Roger Kennedy, "serene and remarkably clean. Its architecture is predominantly rational, not romantic" (1983: xviii).

The Mall: A perhaps even more substantial "cleansing" occurred on the nearby Mall. Envisioned by L'Enfant as a grand boulevard of mansions, museums, fountains, and embassies, the Mall remained largely undeveloped throughout the nineteenth century—or so goes the standard historical narrative, which holds that only in 1901 did the modernizing McMillan Plan "rescue" the area by ridding it of its messy, swamp-like qualities (Wilson 2002, Gutheim and Lee 2006, Savage 2009). A food-centered perspective, however, suggests that the nineteenth-cenury Mall was, in fact, a locus of considerable urban foraging. Indeed, in the capital's earliest years the area served as something of a commons where working people squatted, tended livestock, hunted wild birds, gathered berries and mushrooms for sale at Center Market, and harvested impressive amounts of seafood from the adjacent Canal and Potomac (Smith 1824, Tinckom 1951, Bryan 1914). Even as federal buildings and monuments began to appear, they were surrounded by food production: cattle grazed around the half-built Washington Monument, fish ponds adorned the Botanic Gardens at the foot of Capitol Hill, and the USDA's 1866 headquarters on the Mall (also designed by Cluss, who relished public foodscapes) had working demonstration gardens that yielded seeds for shipment across the country. Commissioned in mid century to turn the wildlands into a respectable public resort, landscape architects Andrew Jackson Downing and Frederick Law Olmstead envisioned a sylvan retreat of pastoral meadows, deer parks, and fruit-bearing trees. All of this gave way after 1901, however, as more modernist planners sought to foster a more austere and ascetic nationalism. As Kirk Savage notes in *Monument Wars*, inviting "public grounds" were replaced by a more abstract, intimidating "public space" (2009: 14). Thus, at the same time that the Archives and Federal Triangle displaced the body—centered Market and Hooker's Division, the Mall's animals, trees, and vegetable gardens were replaced by neo-classical museums surrounded by treeless open spaces, producing a view "as empty as a stretch of western prairie"—but without the cows (Rybczynski 2008: 64). Adjoining the equally sterile Triangle, the modern Mall cut visitors off from local services and residents, particularly African Americans, producing what Lewis Mumford called, "at best a fire barrier, which keeps segregated and apart areas that should in fact be more closely joined" (Savage 2009: 175). Or as former Smithsonian Secretary C. Dillon Ripley observed, his own museum became "someplace you visited after a heavy Sunday dinner" (Kurin 2008: 96). Suitably enough, the whole official zone around the Mall was surrounded by Washington's iconic flowering cherry trees—all form, no fruit. In short, "culture" (in the rarefied, visually "artistic" sense) was materially divorced from cuisine (Korsmeyer 2002).

To be sure, the divorce was never quite complete. Even at the height of modernist asceticism, the Mall and adjacent Ellipse (south of the White House) still hosted a raucous national picnic on July 4—a tradition first established by Thomas Jefferson. Throughout the twentieth century, political demonstrations disrupted the area's official serenity. And starting in the late 1960s, thanks in part to efforts of critics like Mumford and Ripley, the Mall began to offer more bodily delights that appealed to a younger generation awakening to food's commensal and cultural dimensions. First launched in 1967, the Smithsonian Folklife Festival—dubbed the "National Block Party"—brought unusual and purportedly authentic ethnic and regional foods to the heart of the Mall (Kurin 2008). Following up on the Festival's highly popular run during the Bicentennial summer, Jimmy Carter's Inaugural Committee hosted a music and food festival in the same zone during frigid January 1977—a Jacksonian "open house" tradition followed by later Democratic presidents. Soon after, the nearby Old Post Office, built in 1899 right near Hooker's Division and scheduled for demolition as the last stage of Federal Triangle modernization, was renovated and reopened as a headquarters for the National Endowment for the Arts; most telling, its central atrium offered a stylish new food court—a harbinger of several new food services to come on sterile Pennsylvania Avenue. A few blocks away, after frantic lobbying by historic, several other condemned nineteenth relics were converted into museums, hotels, and theaters—the beginning of a major revival that brought trendy restaurants and clubs to D.C.'s crumbling downtown. At about the same time the metropolitan area's food choices were enriched with the arrival of refugees fleeing civil wars (Ethiopia, Vietnam, El Salvador, Afghanistan) and seeking economic mobility (Korea, South America, South Asia) (Cary 1996). Civil rights and eventual home rule (1973) brought new power and respect for D.C.'s exceptionally mobile African American citizens, while young professionals settled in gentrified older neighborhoods made accessible by Washington's deluxe new subway system (1976). With gentrification came still more cafes, bars, farmers markets, street festivals, and hip restaurants. By 2010, with its celebrity chefs (José Andrés), farmers (Joel Salatin), and shoppers (Michelle Obama), D.C.'s dynamic food scene was downright "cool." And yet, as previously stated, this was still framed somewhat ironically, as if a surprise. What was missing was any firm sense of a deeper basis, heritage, or *terroir*. Without this awareness of roots, D.C.'s food offerings seem rather like those of the rest of the affluent world—trendy, volatile, ephemeral, multinational, perhaps temporary. Such stylishness was well-suited to culinary tourism, but it hardly gave local residents a stake in a distinctive place deserving of regard, reciprocity, and responsibility. But can we unearth a more substantial base on which to build?

Imagining/Inventing *Terroir*

Judging from the original French example, place branding entails three components: the identification of a "commensal landscape," entrepreneurial chutzpah, and the creation of compelling founding myths (Hall et al. 2003, Guy 2007, Trubek 2008).

A commensal landscape is an attractive countryside or scene that produces distinctive foods with strong social dimensions (Russo 2009). Conventionally this has usually been associated with vineyards and meadows that are said to yield products (i.e., wine and cheese) with a special, "local" flavor. For the Washington area, the most likely candidate would not be its sparse, depleted farmland but rather its watershed, particularly the Potomac River and its tributaries, Rock Creek and the Anacostia River (Gutheim 1986, Tilp 1978, Wennersten 2008). "Swamp" images aside, for much of the nineteenth-cenury artists viewed D.C. via its rivers, and with good reason, for this landscape was both exceptionally beautiful and rich in food sources (Tinckom 1951). From the earliest European encounters in the seventeenth century through the 1930s, Potomac "fish stories" exulted in the ability of even hapless amateurs to catch astounding amounts of shad, herring, striped bass, oysters, and sturgeon. John Smith established the genre in 1608 when he marveled at "in divers places that abundance of fish, lying so thicke with their heads above the water, as for want of nets (our barge driving amongst them) we attempted to catch them with a frying pan" (Appelbaum 2000: np, Crew 1892: 36). Such accounts fulfilled medieval expectations of Cockaigne, a mythical place of abundance and leisure where animals literally cooked themselves for human pleasure and convenience (Pleij 2003). Similarly in the 1830s the *Niles Weekly Register* noted that it was "not uncommon to pull 4,000 shad or 300,000 herring in one seine haul. One haul of 450 rockfish with an average weight of 60 pounds was documented. Hundreds of sturgeon were captured on a single night near the US Arsenal in Washington" (Cummins 2011: 5). The sturgeon provided caviar— "black gold"—for export from Washington to Europe. Even more extraordinary were the excellent oysters, which were cultivated as close as 40 miles downriver and which formed the basis for an extensive oyster house industry (Tilp 1978). Perhaps the most famous and long lived was Harvey's, which amassed a heap of oyster shells 50 feet high behind its Pennsylvania Avenue location (Evelyn & Dickson 1999). Until fairly recently, oysters were cheap, convenient, and durable. In *Washington By Night* (1936) photographer Volkmar Wentzel noted that "A family would get a barrel of oysters in the fall and keep it out in the garage under damp gunny sacks, feeding the oysters an occasional fistful of cornmeal ..." (1998: 82). But perhaps most memorable were the annual shad runs, as millions of these herring-like creatures made their way up the Potomac to spawn below Great Falls, yielding tons of bony protein and salty roe *en route*. If there was an iconic animal best suited to Washington's identity, it was the lowly shad, an often-despised, bottom-of-the-food chain sojourner fish that tasted best when smoked for inebriated politicians and their minions at annual shad plankings (McPhee 2002). Bringing all of this back from the brink of extinction would, of course, entail an extensive cleansing and regulation of the whole Potomac watershed—but that's precisely one goal of inventing *terroir*, to foster environmental regard, reciprocity, and responsibility.

Terroir is also achieved through the cumulative efforts of brash entrepreneurs who work hard to establish food businesses and invent culinary legacies. Following Krishnendu Ray's study of immigrant restaurateurs, food entrepreneurs serve as the "hinge between taste and toil," and as the "bridge between capital and culture" (Ray 2011). Here, too, D.C. has rich—but often forgotten— traditions upon which to build, starting with the slave gardeners and market hucksters who fed the city's original residents, including President Thomas Jefferson. Most notable was slave Alethea Browning Tanner (c. 1785–1864), whose market garden profits bought her freedom, as well as that of numerous relatives, including John Francis Cook (1810–1855). Cook, in turn, became D.C.'s first black Presbyterian minister, and his descendants became prominent educators, doctors, artists, and civic leaders (Anacostia Museum 2005). Exemplifying the entrepreneurial rags-to-respectability formula outlined by Psyche Williams-Forson in *Building Houses Out of Chicken Legs* (2006), former Big House cooks and waiters served presidents and embassies, opened restaurants, catering firms, hotels, and groceries, and built the foundations of D.C.'s strong black middle class. This often took several generations. For example, blacks had long worked the oyster beds of the region and also sold their harvest on the streets and in rough oyster houses. When Chesapeake-born Thomas Downing (1791–1866) opened a famous oyster house in New York in the 1820s, he was able to finance a private school and elite college education for his son George T. Downing (1819–1903). George in turn harbored fugitive slaves in the restaurant's basement, opened a luxury hotel in Newport, Rhode Island, then moved to D.C. where he ran the members' restaurant in the U.S. House of Representatives for 12 years and became a leading advocate for civil rights and home rule (Ellis 1969: 112, Anacostia Museum 2005: 95, Green 1969: 79). Such stories could apply equally well to more recent generations of immigrant food entrepreneurs, most famous of which was Trinidad-born Ben Ali (1927–2009), who came to D.C. to attend Howard University and then opened a hot dog stand that eventually became the cornerstone of major redevelopment along Washington's U Street, once the center of black cultural life in D.C. As food trucks, ethnic restaurants, street fairs, and farm markets proliferate throughout the area—not just in gentrified urban districts but also in the sterile suburbs and Federal

Triangle—the region seems ready to leverage the "intangible capital" of food-based entrepreneurship into a new wave of Tanner–Downing–Ali success stories.

The third leg of the *terroir* triad is a compelling founding myth. While the entrepreneurial stories may seem inspirational enough, they are not unique to Washington, which is known less for its commercial spirit than for its supercharged political culture. So it seems appropriate to look for a "founding foodie" from the political class (DeWitt 2011). Here the most obvious choice would be Thomas Jefferson, the first president to live full-time in the White House. Jefferson's table was essentially the only game in town during those rude early years, and it was appropriately round to encourage the intimate, informal dinners that he saw essential for creating a new democratic culture. Jefferson's circular table eschewed hierarchical seating, while his use of "dumbwaiters"—plain self-service carts placed around the table—encouraged frank talk without the fear of being overheard by gossiping servants. Occasionally his studied informality might even raise a scandal, as during the "Merry Affair" (1803), when the president refused to escort the new British emissary's wife to the table at a state dinner; instead Jefferson preferred the revolutionary practice of "pele mele"—essentially, rush to the table and grab a seat (Allgor 2000, Scofield 2006). The food itself was French-inspired but served in a generous plantation manner, a deliberate combination of pastoral and cosmopolitan tropes, "republican simplicity ... united with epicurean delicacy" (Kimball 1976: 13). Much has been made of the latter, particularly Jefferson's taste for imported ingredients, dishes, and wines, but Jefferson was equally committed to promoting local products and recipes (Hazelton 1964). His table served Potomac area seafood and game, as well as fresh produce that Jefferson himself purchased from area gardeners—possibly including, in a rather neat union of entrepreneurial and presidential myths, Alethea Browning Tanner (Anacostia Museum 2005: 12). Although Jefferson was a passionate gardener at Monticello, he deliberately restricted the extent of his White House garden so as to encourage local markets, whose weekly offerings he carefully noted in his record books (Leupp 1915: 132, Kimball 1976). And, capping his standing as a booster of horticultural progress, Jefferson gave visitors prized seeds and plants acquired from embassies, much the way a proud cook might send guests home with an extra slice of cake. For example, in one of the most charming and suggestive passages of her rich chronicle of *The Forty Years of Washington Society,* Margaret Bayard Smith remembered a pleasant tea during which "the conversation turned on agriculture, gardening, the differences of both in different countries and the produce of different climates ... Mr. J. gave me some winter melon-seed from Malta" (1906: 50). What could be more appropriate to a town of "sojourners" than attempting to root an Afro-Asian domesticate in local D.C. soil?

Potomac seafood, hustling vendors, Thomas Jefferson—not a bad foundation on which to build a culinary identity.

But, Some Questions

The purpose here has been to extend the logic of *terroir* and locavorism to a place not automatically identified with culinary heritage or distinction. In light of its origins and ongoing function as a neutral common ground belonging to the whole country, and not to its permanent residents, a sense of locality in Washington, D.C. would seem an oxymoron. Deliberately designed to lack political representation (despite high taxation), D.C. has little right to expect much cultural representation either. And yet its very placelessness may render D.C. most representative of a "sojourner" nation. Some have quipped that the U.S. national anthem should be Bruce Springsteen's "Born to Run" (1975), with its memorable lines,

> Baby this town rips the bones from your back
> It's a death trap, it's a suicide rap
> We gotta get out while we're young.

Fifty years ago historian George W. Pierson wrote that the "M-Factor" (mobility, movement, migration) is the essence of Americanness (Pierson 1962). If so then perhaps the Capital's most representative food should be roadside hamburgers and fries suitable for a quick getaway.

The point is not facetious. Who decides a local identity? Can a historian—even one who has lived in the place for almost 40 years—invent one as an academic exercise? Who are the agents of such place branding? All too often traditions and reputations are invented for and by the winners of historical struggles, i.e. the more affluent and powerful (Hobsbawm and Ranger 1983). Such is particularly true in the food business, where distinction and authenticity are so profitably commodified (Johnston and Baumann 2009, Guthman 2011). *Terroir* was first conceived by local French business interests (Boissard 2003, Guy 2007). But does the hospitality industry in D.C., already one of the world's top tourist destinations, even need a culinary identity, as it appears to be profiting quite well without one? Or if local citizens are meant to be the beneficiaries, which citizens? Few doubt the value of reviving the Potomac watershed to its former vibrancy, but will that result in the usual sort of downtown riverfront gentrification that ignores other neighborhoods and denies river access to the poor? (Williams 2001, 2002). Moreover, at a time when many in D.C. are welcoming the arrival of six new Wal-Mart stores, which promise to bring jobs and cheaper food to impoverished sectors, it is not all clear that its most needy residents care much about saving the local institutions and traditions that have failed them so far. Then again, as the world's leading retailer of organic products, Wal-Mart plans to market locally produced food as well (Clifford 2010). The paths to *terroir* are twisted and ironic.

Still they are not necessarily unexpected. Appropriating the deviant and subversive is essential to the hegemonic process by which dominant powers (here: Wal-Mart) retain their hold (Belasco 1989, 2006). Sociologist Todd Gitlin

(1983) has called this the "recombinant strategy" of mass marketers who seek drape their "new" and "modern" products and practices in the comforting guise of the "old" and "traditional"; think how Howard Johnson housed his fast food franchises in neo-classical architecture reminiscent of New England town halls (Belasco 1979). Looking ahead to 2000 from his perch in 1902, shrewd futurist H.G. Wells predicted such post-Fordist, post-modern staging in *Anticipations,* in which he guessed that while most food of the future would be manufactured in "big laboratories" and central kitchens, there would also be a place for the "second-hand archaic" of picturesque "ripe gardens" and anachronistic "craft" products fashioned by "ramshackle Bohemians" inhabiting the "old-fashioned corners" of touristic districts (Wells 1999: 65–109). Similarly, 30 years later British politician Lord Birkenhead envisioned that most food would be produced by industrialized photosynthesis, but thanks to the enduring appeal of "historical romances," "ploughing may even become a fashionable accomplishment, and pig-keeping a charming old-world fancy" (quoted in Belasco 2006: 222). Had they lived into the twenty-first century, neither Wells nor Birkenhead would have been surprised by Wal-Mart's cooptation of the organic and local.

Those committed to the more subversive implications of *terroir,* especially the three Rs, may need to build their insurgency around iconic products that are simply not co-optable, even by Wal-Mart. Shadburgers anyone? For D.C. that just might work.

References

Adams, H. 1889. *A History of the United States During the First Administration of Thomas Jefferson. Vol.1.* New York: Charles Scribner's Sons.

Allgor, C. 2000. *Parlor Politics: In Which the Ladies of Washington Help Build a City and a Government.* Charlottesville: University of Virginia Press.

Anacostia Museum. 2005. *The Black Washingtonians.* Hoboken, NJ: Wiley.

Appelbaum, R. 2000. John Smith's Fish: Mapping Natural Resources, Cultural Habits, and Food. *Texts of Imagination and Empire.* Retrieved June 10, 2009 at http://www.folger.edu/html/folger_institute/jamestown/c_appelbaum.htm.

Bednar, M. 2006. *L'Enfant's Legacy: Public Open Spaces in Washington, D.C.* Baltimore: Johns Hopkins University Press.

Belasco, W. 1979. Toward a Culinary Common Denominator: The Rise of Howard Johnson's. 1925–1940. *Journal of American Culture,* 2: 3, 503–18.

Belasco, W. 1989. *Appetite for Change: How the Counterculture Took on the Food Industry.* New York: Pantheon Books.

Belasco, W. 2006. *Meals to Come: A History of the Future of Food.* Berkeley: University of California Press.

Belgacafe.com. 2005. Metro Washington's top restaurants and chefs revealed at "RAMMY" Awards announcement event. Retrieved November 21, 2011 from http://www.belgacafe.com/uploads/20050328.pdf.

Boissard, P. 2003. *Camembert: A National Myth*. Berkeley: University of California Press.

Brown, L.W. 1972. *Free Negroes in the District of Columbia, 1790–1846*. New York: Oxford University Press.

Bryan, W.B. 1914. *A History of the National Capital. Vol.1, 1790–1814*. New York: Macmillan.

Bryan, W.B. 1916. *A History of the National Capital. Vol. 2, 1815–1878*. New York: Macmillan.

Carson, B. 1990. *Ambitious Appetites: Dining, Behavior, and Patterns of Consumption in Federal Washington*. Washington: American Institute of Architects Press.

Cary, F.C. 1996. *Urban Odyssey: A Multicultural History of Washington. D.C.* Washington: Smithsonian Institution Press.

DeWitt, D. 2010. *The Founding Foodies: How Washington, Jefferson, and Franklin Revolutionized American Cuisine*. Naperville IL: Sourcebooks.

Clifford, S. 2010, October 14. Wal-Mart to Buy More Local Produce. *New York Times*. Retrieved on November 22, 2011 from http://www.nytimes.com/2010/10/15/business/15walmart.html.

Crew, H.W. 1892. *A Centennial History of the City of Washington*. Dayton, OH: United Brethren Publishing House.

Cummins, J. A Compilation of Historical Perspectives on the Natural History and Abundance of American Shad and Other Herring in the Potomac River. Retrieved November 22, 2011 from http://www.potomacriver.org/cms/wildlifedocs/shadhistory032011.pdf.

Dickens, C. 1842. *American Notes for General Circulation. Vol. 1*. London: Chapman Hall.

Ellis, J.B. 1869. *The Sights and Secrets of the National Capital: A Work Descriptive of Washington City in All Is Various Phases*. New York: United States Publishing Co.

Evelyn, D.E. and Dickson, P. 1999. *On This Spot: Washington, D.C.* Washington, D.C. National Geographic.

Gatewood, W.B. 2000. *Aristocrats of Color: The Black Elite, 1880–1920*. Fayetteville: University of Arkansas Press.

Genovese, E.D. 1976. *Roll, Jordan, Roll. The World the Slaves Made*. New York: Vintage.

Gitlin, T. 1983. *Inside Prime Time*. New York: Pantheon Books.

Goode, J.W. 2003. *Capital Losses: A Cultural History of Washington's Destroyed Buildings*. Washington: Smithsonian Books.

Green, C.M. 1962. *Washington: A History of the Capital, 1800–1950*. Princeton: Princeton University Press.

Green, C.M. 1969. *Secret City: History of Race Relations in the Nation's Capital*. Princeton: Princeton University Press.

Gutheim, F. 1986. *The Potomac*. Baltimore: Johns Hopkins University Press.

Gutheim, F. and Lee, A.J. 2006. *Worthy of the Nation: Washington, D.C. from L'Enfant to the National Capital Planning Commission.* Baltimore: Johns Hopkins University Press.

Guthman, J. 2011. *Weighing In. Obesity, Food Justice, and the Limits of Capitalism.* Berkeley: University of California Press.

Guy, K.M. 2007. *When Champagne Became French: Wine and the Making of a National Identity.* Baltimore: Johns Hopkins University Press.

Haley, A.P. 2011. *Turning the Tables: Restaurants and the Rise of the American Middle Class, 1880–1920.* Chapel Hill: University of North Carolina Press.

Hall, C.M., Sharples, L., Mitchell, R., Macionis, N. and Cambourne, B. 2003. *Food Tourism Around the World.* Oxford: Elsevier.

Hazelton, J.H. 1964. Thomas Jefferson, Gourmet. *American Heritage Magazine.* 15: 6. Retrieved August 2, 2010 from www.AmericanHeritage.com.

Hobsbawm, E. and Ranger, T. 1983. *The Invention of Tradition.* Cambridge: Cambridge University Press.

Jacob, K.A. 1993. *Capital Elites: High Society in Washington, D.C. after the Civil War.* Washington, D.C.: Smithsonian Institution Press.

Johnston, J. and Baumann, S. 2009. *Foodies: Democracy and Distinction in the Gourmet Foodscape.* New York: Routledge.

Kennedy, R. 1983. Introduction to *The WPA Guide to Washington, D.C.* New York: Pantheon Books, xvii–xxx.

Kimball, M. 1976. *Thomas Jefferson's Cook Book.* Charlottesville: University Press of Virginia.

Korsmeyer, C. 2002. *Making Sense of Taste: Food and Philosophy.* Ithaca: Cornell University Press.

Kurin, R. 2008. Culture of, by, and for the People: The Smithsonian Folklife Festival. In N. Glazer and C.R. Field (eds), *The National Mall: Rethinking Washington's Monumental Core.* Baltimore: Johns Hopkins University Press, 93–113.

Landau, B.H. 2007. *The President's Table: Two Hundred Years of Dining and Diplomacy.* New York: HarperCollins.

Lessoff, A. 1994. *The Nation and Its City: Politics, Corruption and Progress in Washington, D.C., 1861–1902.* Baltimore: Johns Hopkins University Press.

Leupp, F.E. 1915. *Walks About Washington.* Boston: Little, Brown.

Luria, S. 2006. *Capital Speculations: Writing and Building Washington, D.C.* Lebanon NH: University of New Hampshire Press.

McPhee, J. 2002. *The Founding Fish.* New York; Farrar, Straus, Giroux.

Miller, B. 1990, September 24. Dining In the Capital: It's a New Deal. *New York Times.* Retrieved November 21, 2011, from http://www.nytimes.com/1990/01/24/garden/dining-in-the-capital-it-s-a-new-deal.html?pagewanted=all&src=pm.

Miller, D.C. 1989. *Dark Eden: The Swamp in 19th century American Culture.* Cambridge: Cambridge University Press.

National Museum of the American Indian Site. 1997. Archaeological Investigations. Retrieved November 26, 2011 from http://www.si.edu/oahp/nmaidig/#1997.

Pierson, G.W. 1962. The M-Factor in American History. *American Quarterly.* 14 (2): 275–89.

Pleij, H. 2003. *Dreaming of Cockaigne: Medieval Fantasies of the Perfect Life.* New York: Columbia University Press.

Provine, D. 1973.The Economic Position of the Free Blacks in the District of Columbia, 1800–1860. *The Journal of Negro History.* 58(1): 61–72.

Ray, K. 2011. Dreams of Pakistani Grill and Vada Pao in Manhattan. *Food, Culture and Society.* 14(2): 243–74.

Rice, J.D. 2009. *Nature and History in the Potomac Country: From Hunter-Gatherers to the Age of Jefferson.* Baltimore: Johns Hopkins University Press.

Russo, R.A. 2009. *Using a Socio-Cultural Framework to Evaluate Farmland Preservation Policy Success in Maryland.* Unpublished doctoral dissertation. University of Maryland, College Park.

Rybczynski, W. 2008. "A Simple Space of Turf": Frederick Law Olmsted Jr.'s Idea for the Mall. In N. Glazer and C.R. Field (eds), *The National Mall: Rethinking Washington's Monumental Core.* Baltimore: Johns Hopkins University Press, 55–65

Savage, K. 2009. *Monument Wars: Washington, D.C., The National Mall, and the Transformation of the Memorial Landscape.* Berkeley: University of California Press.

Scofield, M.E. 2006. The Fatigues of His Table: The Politics of Presidential Dining During the Jefferson Administration. *Journal of the Early Republic.* 26: 449–69.

Seifert, D.J. 1991. Within Sight of the White House: The Archaeology of Working Women. *Historical Archaeology.* 25(4): 82–108.

Sietsema, T. n.d. Washington, D.C.'s Best Restaurants With the help of several talented young chefs, the nation's capital has become a fresh and lively dining destination. Epicurious.com. Retrieved November 21, 2011 from http://www.epicurious.com/articlesguides/diningtravel/restaurants/washington#ixzz1eMKaTdgm.

Smith, M.B. 1824. *A Winter in Washington; or Memoirs of the Seymour Family. Vol.1.* New York: E. Bliss and E. White.

Smith, M.B. 1906. *The First Forty Years of Washington Society.* New York: Charles Scribner's Sons.

Strauss, D. 2011. *Setting the Table for Julia Child: Gourmet Dining in America, 1934–1961.* Baltimore: Johns Hopkins University Press.

Tangires, H. 1999. Meeting on Common Ground: Public Markets and Civic Culture in Nineteenth Century America. Unpublished doctoral dissertation. George Washington University, Washington, D.C.

Tangires, H. 2008. *Public Markets.* New York: Norton.

Tilp, F. 1978. *This Was Potomac River.* Washington: Tilp.

Tinckom, M.B. 1951. Caviar along the Potomac: Sir Augustus John Foster's "Notes on the United States," 1804–1812. *William and Mary Quarterly.* 8(1): 68–107.

Trubek, A. 2008. *The Taste of Place. A Cultural Journey into Terroir.* Berkeley: University of California Press.

Wells, H.G. 1999 [1902]. *Anticipations.* Mineola, NY: Dover Publications.

Wennersten, J.R. 2001. *The Chesapeake: An Environmental Biography.* Baltimore: Maryland Historical Society.

Wennersten, J.R. 2008. *Anacostia: The Death and Life of an American River.* Baltimore: Chesapeake Books.

Wentzel, V. 1998. *Washington by Night: Vintage Photographs from the 30s.* Golden, CO: Fulcrum Publishing.

Williams, B. 2001. A River Runs Through Us. *American Anthropologist.* 103(2): 409–31.

Williams, B. 2002. Gentrifying Water and Selling Jim Crow. *Urban Anthropology and Studies of Cultural Systems and World Economic Development.* 31(1): 93–121.

Williams-Forson, P.A. 2006. *Building Houses Out of Chicken Legs.* Chapel Hill: University of North Carolina Press.

Wilson, R.G. 2002. High Noon on the Mall: Modernism vs. Traditionalism, 1910–1970. In R. Longstreth (ed.), *The Mall in Washington.* New Haven: Yale University Press, 143–67.

Young, J.S. 1966. *The Washington Community, 1800–1828.* New York: Harcourt Brace.

Chapter 3

Of Cheese and Ecomuseums: Food as Cultural Heritage in the Northern Italian Alps[1]

Cristina Grasseni

In Italy, the concept of cultural heritage is increasingly being applied to local food in the name of its "typicity." The notion of a food being "typical" is inextricably tied to that of its being local, namely rooted in a specific territory with its biological and botanical peculiarity. Typical foods are not just "traditional" but are bearers of the historical know-how that would flourish in that particular locality. Similarly to the idea of "*terroir*," the notion of typicity thrives in connection with the capacity of trade networks and supply channels to guarantee and protect geographical indications and denominations of origin (Grasseni 2012d).[2]

This paper offers two examples of how the protection of food as cultural heritage may not be disjoint from a degree of "invention" (Hobsbawm and Ranger 1983) or even "reinvention" (Grasseni 2007). In particular, I shall highlight the link between typical products and the food industry as well as the role played by regional and national institutions in supporting the concept of food "typicity" and its value as a form of cultural heritage and territorial patrimony.

The relationship between the protection of geographical indications and strategies of regional development will become apparent especially with reference to relatively new cultural institutions such as the Ecomuseum, which in northern Italy has so far tended to play the card of traditional continuity whilst glossing over technological change in local food production.

On the one hand, the "reinvention of food" (Grasseni, 2007) rests on newly rediscovered cultures of taste and on a notion of *tipicity* that implies an intimate tie between local foodstuffs and local territories. Several "technologies of localization" are firmly in place to standardize environments, *routines* and protocols of production whilst geographical indication legislation calibrates local products to a global market. On the other hand, the ritualization of consumption

---

1   An earlier version of this work appeared in French as Grasseni 2012d.

2   Food "typicality" was the focus of a recent Conference at Parma University's Food Lab in September 2010 (Grasseni, 2013) as well as of a Colloque at the University of Nîmes in February 2010 on cultural patrimony and locality (Grasseni, 2012d). This work owes much to the discussion and conversations initiated on those occasions.

(through media exposure, for instance) adds to what I would call a virtualization of local foods. By this I mean that local foods that acquire the status of "typical" foodstuffs become sought-after commodities, which are consumed because of their alleged pedigrees (such as a long historical heritage or direct link to a specific landscape, or a specific technique for producing, seasoning, maturing or shaping the produce). In other words, pedigree qualities are additional and not reducible to the inherent and immediately perceivable qualities of food such as taste, smell, texture or appearance. As a result of this, in TV programs, festivals and publicity, local foods are ever more depicted as icons of specific territories, through reference to *terroir* and local identity. They are consequently treated as political and economic resources. To claim "typical" status for local foodstuffs not only means re-inventing them (just as "traditions" are invented or constructed[3]): it can mobilize strategies of self-rediscovery of patrimonialization of local histories, places and landscapes as a form of "intangible" goods.

Since the 1970s, the ambivalent relationship between "local cultures and their commodification" was an object of anthropological critique (Lombardi Satriani 1973), as it is today in the face of global agendas of heritagization (Hemme et al. 2007). Alpine ecomuseums investing in local foodstuffs provide a complex and novel ethnographic context to review the commodification of food as heritage that was hitherto scarcely explored. Ecomuseums should not be misconstrued as heritage or natural theme parks. The latter follow the model of the Skansen open air museum[4] and foster cultural and environmental tourism through the reenactment of traditional activities with period objects and architectures, according to the peculiar styles of specific provinces and territories. Against the idea of a standardized and high-brow presentation of cultural and natural heritage, ecomuseums were ideated by the revolutionary museum critics of the *nouvelle museologie* such as George Henry Riviére.[5] One of its contemporary advocates and inventors, Hugues de Varine, underlines that ecomuseums are not focused on the past (unlike traditional museums), neither are they eco-parks focusing on natural conservation, but should be the manifestation of a collective local project. "Community museums" are animated by a local action group that determines how to define local patrimony (whether natural, material, or immaterial) through the

---

3 Hobsbawm and Ranger's initially iconoclastic idea (1983) is now largely accepted in current scholarship. See nevertheless Regina Bendix's critique of their distinction between authentic and invented tradition (1997). My own thesis is that in food heritage tradition is necessarily re-invented. It is namely rephrased and re-designed within the current frame of patrimonialization, commodification and accountability of local foodstuffs (see Grasseni 2011).

4 Skansen, the first example of an open air theme museum, was opened in Stockholm in 1891, featuring detailed reconstructions of Scandinavian rural life, which at the time was perceived to be disappearing in the face of industrial modernization.

5 Rivière (1992). For an informative overview on the debates of that period and a selection of original texts, in French, see Desvallées (1992).

activities of "interpretation centres." Ecomuseums are therefore very difficult to define in general. Their development strategies may include tourism, but also solidarity economy networks, or commercial strategies focusing on local products (de Varine 2005).

The ecomuseum projects that I present here were actually founded with local and regional institutional support, and acknowledge a strategic convergence with local commercial strategies aiming at casting local foodstuffs as "typical" in order to capitalize on their patrimonialization as local heritage. Their agendas range from the defense of traditional cheese recipes in the face of standardization (as in the *Ecomusei delle Valli del Bitto*, supporting the local Slow Food Bitto cheese presidium), to the rediscovery of lost recipes as a strategy to identify a distinctive locality with a specific local product (as in the *Ecomuseo Val Taleggio: Civiltà del Taleggio, dello Strachitunt e delle Baite Tipiche*). I shall discuss the relationship between food ecomuseums and local economies on the basis of these two examples.

The Ecomuseums of the Valli del Bitto

The first example I am proposing hinges on the controversy surrounding the authenticity of Bitto cheese. The rising demand in the 1990s for *Bitto*, a renowned *alpage* cheese produced in the Lombard Alps north of Milan, resulted in a progressive expansion of the area producing Bitto cheese according to its geographical indication.[6] Boundary-marking being a key element in the social construction of food as heritage, the hotly debated issue is whether Bitto cheese is a patrimony of the Bitto valleys *only*, or of the larger province of Sondrio in northern Italy (Grasseni 2012b). This issue played out as a conflict of interests between the Bitto valleys on the one side and a wider sphere of economic and political interests involving among others the agricultural trade union and the PDO consortium (Grasseni 2012c).

The general expectation is that obtaining a geographical indication such as a Protected Designation of Origin (PDO) for local foodstuff should not imply standardization but rather distinction from standard production. In some cases though, as that of the Bitto Cheese, a small number of local artisan cheese-makers maintain that the PDO Bitto cheese is more standardized and lower quality than the authentically traditional Bitto cheese. The "rebel" cheese-makers, backed up by Slow Food, produce their traditional recipe outside and against the PDO Bitto consortium—and as a result they cannot name it "Bitto" (Grasseni 2012a).

If we think of "patrimony" in terms of a *diversity* of the cultural environment, as well as of the biological and natural environment (an idea also explored by

---

6   A detailed though avowedly partisan history of such enlargement is provided by Corti and Ruffoni (2009), who defends the case of the *Associazione Produttori Valli del Bitto* against the PDO consortium on a number of historical and technical grounds.

Di Giovine 2008), it is easy to recognize how cheese-making practices entail a vast amount of diversified local knowledge that is directly connected to specific issues of management of territory and landscape. Implicit and explicit references to continuity and non-contamination add to the appeal of the Alps as bastions of "typical" foods. Such were the cultural references, for instance, successfully used to obtain EU "Leader" funds in several European marginal rural areas,[7] especially in connection with "age-old" practices such as *alpage* cheese-making and transhumant pastoralism.[8]

As in the case of the Bitto cheese, the process of patrimonialization of such diversity may have ambivalent outcomes, especially as individual cheese-makers hardly play a role in making and changing rules and procedures regarding PDO protocols. In the Bitto case, these decisions regarded the extension of the area of production (which was extended beyond the Bitto valleys), the diet of the cows (which came to include fodders in addition to hay and grass) and the types of milk used for producing Bitto cheese (which changed the goat/cow milk ratio in favor of cow milk). The small producers of the Bitto valleys felt that they were not sufficiently involved, even though they were initially all members of the consortium.

Several social actors contribute to the status of Bitto as heritage cheese, and journalists play an important part. After a Milan conference on the Bitto *vexata quaestio*, on April 25, 2009, Enzo Lo Scalzo, ASA-press delegate for Lombardy and culinary expert, wrote that Bitto cheese was already being used in the 1960s as an ingredient for delicate local recipes including local *Sfursat* wine—seasoned in chestnut barrels—salami and pickles—and defined it as a "cheese for meditation and hospitality."[9] The readers of Stefano Mariotti's online magazine *Quale Formaggio* (Which Cheese?—formerly *Cheese Time*) acknowledged on the web the filling capacity of this well matured cheese, along with the persistence of its aroma and the diversity in taste between Bitto cheeses made on different pastures.[10]

7　Leader+ is a rural development initiative funded by EU structural funds and designed to integrate rural actors in long-term local and regional networks: http://ec.europa.eu/agriculture/rur/leaderplus/index_en.htm. It ran under three different plans (Leader, Leader 2 and Leader Plus) until 2006 with uneven impact on local communities.

8　By transhumance I refer here to the equivalent of the French *alpage* practice of moving the herds seasonally between winter and summer pastures. Taking herds uphill in the summer to graze higher pastures frees the village pastures from cattle, so that hay can be made and be stored away for the winter (McNetting 1981). Transhumant trails could extend over hundreds of kilometers and entail a longer period than just the summer *alpage*. Nevertheless, in Italy the word "transhumant" is still used for what are by now short summer trips to the closest higher pastures. On the distinction between *alpage* and transhumance see Jones (2005).

9　Associazione Stampa Agroalimentare, http://www.asa-press.com.

10　Fabio Tarpanelli, April 24th, 2009 06:33, http://www.qualeformaggio.it/index.php?option=com_content&view=article&id=211:proseguono-le-iniziative-a-favore-del-bitto-storico&catid=4:attualita&Itemid=10.

Both testimonials hark back to a time of virtuousness and of meaningful scarcity, sobriety, and spiritual wholeness as well as social integration. This is by no means uncommon in the collective representations of "authentic" alpine cheese, as can be seen with the rediscovery of a number of local artisan cheese traditions, from *Stracchino* to *Strachitunt* (to which I return in the next section). As a result of this, the recently established Ecomuseums of the Valli del Bitto could not help but play an ancillary role in the Bitto controversy.

Both municipalities of the Bitto Valleys (Albaredo and Gerola Alta) have quickly seized the institutional and economic opportunity of an innovative law approved by the Region Lombardy in 2007, to be officially recognized as "Ecomuseum of the Bitto Valleys" and as "Ecomuseum of Val Gerola." If we read the description of the Bitto Valleys Ecomuseum, we are told that

> the territory of the Valle del Bitto di Albaredo is a landscape where the natural component was forged by the patient and headstrong skill of the local people who, during centuries of cohabitation established a tacit compromise with the natural elements. The cultural landscape which ensued is mostly made of knowledge, experiential wisdom and orally handed down practices. The relinquishment of traditional mountain lifestyles is the symptom of the fact that these practices have been interrupted and those skills forgotten. This is due both to the fact that traditional animal husbandry is uneconomical, and to the deeply rooted conviction that farming in the mountains is useless. The mountain peasant's self perception is, in fact, that of coming last in the ladder of the economic world .... Nevertheless today's market allows for solutions to the damage caused: sustainable rural tourism that rediscovers traditions and taste.[11]

Diversity is described here as an endangered resource. The underlying equation establishes the ecomuseum as a tool to attract cultural tourism and to commercialize local cheese, as this would provide an economically viable path to cultural and environmental resilience. As a symptom of the name-war between the Bitto PDO consortium and the rebel cheese-makers, which has involved the municipalities on either side of the controversy, Bitto is in this Ecomuseum's name but Bitto cheese nevertheless shines for its absence from the list of the local patrimony that is implicitly encased in the brochure: landscape, skills, and traditional mountain lifestyles. Elsewhere in the website, one can find a list of available cheeses for sale (including Bitto PDO as well as other less renowned local cheeses such as *matusc*). The Ecomuseum here becomes what common sense would have it: an open air park, complete with theme routes (the route of the furnace, of the dairy, and of charcoal-making). It provides an ecological packaging of cultural and environmental diversity for advertising the valley's produce. The coop dairy *Caseificio AlpiBitto* inaugurated in March 2006 is described as a keystone

---

11   From http://www.vallidelbitto.it/ecomuseo_valledelbitto.html. My own translation from the Italian text.

producer of 2,500 kilograms of cheese a month, transforming local milk ("on average 800 liters a day just from the two main village breeders, plus the other smaller enterprises"). The dairy resides on a multi-function development lot hosting a restaurant, a bakery and a call center, "Alps' Word."

The Slow Food Presidium and the "Bitto Center" (an unauthorized denomination since it sells non-PDO "bitto" cheese) are instead part and parcel of a competing Ecomuseum, also officially acknowledged by the regional government, set up by the Municipality of Gerola Alta as "Ecomuseum of Val Gerola." The banner *Classic Bitto from the Historical Pastures* ("Il Bitto classico degli alpeggi storici") hanging from the walls of the Valgerola Ecomuseum's Bitto Center, was contested by the inspectors of the Ministry for Agriculture in October 2009 and had to be taken down. Faced by regulatory censorship, the Slow Food support was of paramount importance to the traditional producers, granting them visibility and customers, as well as moral authority, and guaranteeing them a dedicated niche of customers and tourists. The website of the municipality of Gerola Alta unmistakably links the Ecomuseum Val Gerola with the rebel cheese-makers, the *Associazione Produttori Valli del Bitto*, and their website http://www.bittocheese.com/, in English as well as Italian, dedicated to "Heritage Bitto." Their Bitto Center is described as "a building that aims to be a reference point for the promotion of the entire valley," comprising a maturing department, an outlet for non-PDO Bitto cheese and the seat of the Ecomuseum. It is openly acknowledged that "the center's management is in the hands of a company that is expression of the local producers' association, who are also involved in the Ecomuseum together with other local associations such as the Folklore group, the tourist board, the parish, the Alpine Corps Association, sports groups and the village choir."[12] This Ecomuseum's "Typical" products obviously enlist Bitto cheese with other local dairies such as *mascherpa*.[13]

This confrontation between two ecomuseums in two neighboring villages is part of the emergence of novel actors and institutions that aim at establishing their own interpretation of foodways as heritage and patrimony—and correspondingly their commercial and political power. The Bitto controversy shows how social actors articulately claim the notion of heritage: as diversity, distinction, lineage, patrimony, and finally political power. Communication plays a crucial role and this is also reflected by the symbolic special treatment that the champions of Heritage Bitto receive at the Bitto Center: in the maturing department there's a special section devoted to the 12kg, 10-year-old cheeses that carry a branded dedication to the Slow Food spokesmen and gourmet journalists who publicly defended the rebel producers' association.

---

12   http://www.valgerolaonline.it/ecomuseo/ecomuseo3.html.
13   http://www.ecomuseovalgerola.it/05_prodotti/01_prodotti.html.

The Case of *Taleggio* and *Strachitunt*

Strachitunt and *Taleggio* cheese from Val Taleggio, in northern Italy, represent two different strategies of heritigization that are in fact of paramount importance to the fate of the valley and its inhabitants. *Taleggio* is well known in Italy and abroad as a soft, fresh white square cheese which is commercialized after being aged for up to three months. *Taleggio* is protected by a PDO (protected designation of origin). It is an unpretentious cheese for daily consumption and owes its name to the valley of Taleggio (Val Taleggio) in the province of Bergamo. Though currently technically different from the fresher, fatter cheese "*stracchino*," which is not protected by a geographical denomination, there is historical evidence that *taleggio* is the denomination that the *stracchino* cheese of the Taleggio valley has assumed, since the beginning of the nineteenth century, after winning international recognition as "*stracchino di Taleggio*" (*stracchino* cheese from the Taleggio valley).

The well-known *Inchiesta Jacini*, which commented on the state of Italian agriculture upon mandate of the newly unified national government in 1881–86, stated that "the noun *stracchino* ... stems out of the soft cheeses that *transhumant herders* produce hastily during their transit between mountain and lowland and back, while resting on the way, with the milk of the cows *worn out (stracche)* from the long journey" (Jacini 1883 vol. 6 tomo1 fasc. 1). However hastily, humbly, and tiredly produced, the *stracchino* of Taleggio won the gold medal at Paris' International exhibition in 1909.[14] The name *taleggio* was recorded by homonymy with Val Taleggio in 1918, in Alfredo Panzini's *Dizionario Moderno* (Stefanelli, 1998).

In 1934 a "*consorzio produttori stracchini di taleggio*" was constituted with an act registered by notary public Antonio Leidi, naming 41 members—all resident within the valley—as producers (*soci produttori*). Nevertheless, plagued by outmigration, the valley's economy could not capitalize on its homonymy with a well known and widely sold cheese.[15] Only in 2006 did the local cooperative of producers start branding their cheese *stracchino di Vedeseta*, to distinguish their raw-milk production from the standard PDO *Taleggio* produced from pasteurized milk. They also tried to brand it in dialect, as *strachì quader*, or as *stracchino di Taleggio* to claim the distinction and the historical origin of the *stracchino* cheese from Val Taleggio. Nevertheless the consortium for the protection of the geographical indication of *Taleggio* cheese forbade using the name "*taleggio*" in any other way than the PDO prescribed.

The problem is that *Taleggio* cheese, although bearing the name of the valley, has a PDO production area so vast that it is produced and sold industrially at low prices throughout northern Italy—hardly serving its auspicious function of a

---

14   The original certificate is displayed in a frame in the home of Taleggio's retired teacher, whose grandfather was a cheese merchant.

15   The total resident population of Val Taleggio is now less than 800, whereas it exceeded 5,000 in the sixteenth century.

"typical" product that might revive the economy of Val Taleggio. The historical reason for this is that *taleggio* cheese was produced during transhumance with a simple recipe and a short maturation time, so that over the centuries it spread in the northern Italian lowlands and became effectively staple food. This may well be a source of pride for Val Taleggio, but it impedes its economic exploitation as heritage food. Several forms of recognition of this moral and historical pre-eminence were demanded by the valley's producers: they asked to have the seat of the consortium established in the valley (with corresponding job opportunities for the local youth), or to have a royalty paid from the consortium to the valley's "historical producers." But a valley producing only about one percent of the actual industrial production of *taleggio* cheese is hardly well placed to obtain special recognition within the PDO consortium.

The local producers then pinned their hopes on a commercially successful but strictly local product, *Strachitunt*, which could increase visibility for all of the valley's cheese (Grasseni 2012c). The skills needed to produce *Strachitunt* had almost been lost, until a local cheese merchant invested in a publicity campaign to save *Strachitunt* from extinction in the late nineties. This is a Gorgonzola-type of cheese, made by mixing the curd from two consecutive days in such a way that, during maturation, natural blue moulds develop. In other words, this is a natural blue cheese from unpasteurised milk that develops a sharp combination of hot and sweet flavors, thanks to highly diversified moulds.

Strachitunt being much rarer and peculiar in taste and texture than the widely commercialized *taleggio* cheese, the Val Taleggio administration and producers set out on obtaining a PDO for it, to brand it as "the valley's cheese." A consortium of cow breeders and cheese merchants requested a DOP for the valley's *Strachitunt* in 2004, and an Ecomuseum (tellingly named "Civilization of *Taleggio* cheese, of Strachitunt and of its typical mountain abodes")[16] was established in support of sustainable tourism and of the local cooperative of production. The Turin Slow Food Salon exhibited both *taleggio* and *strachitunt* in 2004 (Grasseni 2012b). This public presentation was part of a negotiation to establish a Slow Food presidium in Val Taleggio. Unfortunately, this process was not successful because the producers did not accept to bear the costs of re-converting to organic farming (thus avoiding fodder, dietary supplements and imported hay for the production of *Strachitunt*).

The PDO certification on the other hand proved lengthy and laborious, as cheese producers and traders from outside the valley claimed that they knew the recipe for *Strachitunt* and had the right to interpret it and produce it outside the valley—it was, after all, a transhumant tradition, so a very mobile one! The Ecomuseum was established to bolster, showcase and coordinate the momentous transformations that Val Taleggio was undergoing: *Strachitunt* had been reinvented but had not been sufficiently protected as food heritage, either by a Slow Food Presidium or by a PDO. As a result the valley's producers felt exposed

---

16 *Ecomuseo Val Taleggio—Civiltà del Taleggio, dello Strachitunt e delle Baite tipiche.*

to unfair competition from bigger producers who could sell a cheaper product with "their" name (again!). This mood was palpable at the public audience of the case for a protected denomination of *Strachitunt* on October 29, 2010, an official milestone setting the ground for the final recognition of *Strachitunt*'s geographical indication. After the positive evaluation by the Regional government published in June 2009,[17] the national Ministry of Agriculture set a date for a public hearing.

In a crowded municipal hall in the 250 inhabitants village of Vedeseta, the president of the consortium, the main cheese merchant of the valley and currently president of the Ecomuseum of Val Taleggio, recognized the long and laborious nature of the normative process but also of the relational context in which it evolved over six years, thanking the local administrations but reminding them that this is not the project of a single producer or cooperative but a strategic and participated plan for the valley: "this product exists, its history is true, the tradition is there. Those who produce it are there, those who will produce it in the future are here. Milk is here, quantity is immaterial. We wish to make our own produce according to our own capacities and competences. We obviously want to develop it in the sense of tourism, restaurants, and breeding improvement."[18] After this, he read out a fax received at 9.39 on that very morning from the neighboring mountain administration (*Comunità Montana*) on behalf of the cooperative of producers of the province of Lecco, to ask for an enlargement of the production area to that province. The president of the consortium noted that these administrators had been informed about the *Strachitunt* project over the previous seven years and lamented the untimeliness and opportunism of this request, well knowing that it would constitute a further obstacle and slowing down of the denomination. The functionary of the Ministry for Agriculture and of the Region Lombardy then followed, with the public reading of the protocol for production as proposed by the consortium, asking everyone present to pick up a copy and to sign the attendance sheet. Once publicly verified, the protocol proposal would be published in the official gazette bearing all Italian laws, the *Gazzetta Ufficiale*, pending 30 days during which any contrasting observation could be raised by any Italian citizen. Following that, the national government would propose the denomination for registration to the European Commission. Once registered, this voluntary "disciplinary protocol" becomes binding by law for anyone wishing to brand their produce as "*Strachitunt.*" The long-awaited national decree was published on January 21, 2011, but in February 2011, just before the expiry of the 30 days, the official recourse by three large producers of the Bergamasque plains came, requesting that the area of production be extended to the province of Bergamo, thus further postponing the official recognition of the brand at least within Italy, whilst waiting for the final European stamp. Well after that, in February 2012, backed by historical documentation that proves historicity and continuity of production in the

---

17   *Decreto della Direzione Generale Agricoltura* n. 5101 del 22/05/2009, BURL n. 23—Serie Ordinaria—8/06/2009.

18   Vedeseta (BG), 29 October 2010, Public Audition.

mountains and not in the lowlands, a final and conclusive audition in Rome turned down the exceptions of the lowland producers and accepted the documentation of the mountain producers, all of whom were present in person.

In the meantime, I was observing my mountain cheese-making friends as they learned to cast themselves as living heritage. In particular the last cow breeder who had been preserving the recipe through dwindling production, appeared on local, then national media as "savior" of their authentic traditional mountain cheese. Scaling up the family business in fact went hand in hand with playing a public persona as the authentic peasant in the heritage scenario, as prominent testimonials of the Ecomuseum Val Taleggio at the Slow Food *Salone del Gusto*, the biennial food fair held in Turin. There, in October 2010 I found myself alone selling *Strachitunt* cheese at their stand, while my friends flocked to see the video-installation of the Ecomuseum Val Taleggio next door, featuring their portraits projected on a screen of fake milk in a copper cauldron. This cinematographic climax confirmed in fact the emergence of a new local elite, a single family of breeders and cheese-makers winning all the prizes at the valley's cattle fair, becoming president of the producers' co-operative, and eventually renting the abandoned pastures of the neighbors whose heirs had left the village.

Conclusion

I have presented two cases of mountain cheeses whose reinvention resulted in heightened local competition, and I have focused in particular on how this confrontation reverberates in significant ways on the social and institutional contexts of the actors involved. The "name war" between PDO Bitto versus Heritage Bitto involves not only the cheese-makers of the PDO consortium and the *Associazione Produttori Valli del Bitto* but also local administrations with their respective ecomuseums, Slow Food activists, and food journalists. The articulation of food heritage as "typicity" and "diversity" brought a positive result to the producers of Heritage Bitto, who used communication and networking in the face of normative censorship. In the second case, the re-invention of *Strachitunt* added pressure on Val Taleggio's weakened local system of production, notwithstanding the effort made by the valley's Ecomuseum in showcasing their heritage cheese.

Both stories underline several aspects of the process of heritagization: the issue of authenticity, the demand for novel communicative and institutional competences form local actors; and the need for political support to ensure commercial success.

As Michael Herzfeld has shown, in the demise of artisanal culture, diversity becomes a commodity and, in order to be acknowledged as such, tradition must be translated into a form of patrimony. Food-oriented ecomuseums can combine "immaterial resources" and "material culture" to be marketed in the form of "life-experiences"—ecomuseums being easier to market than labeling products. The paradox of current attempts to reinvent tradition and add value to local products lies in the impact of such "global hierarchies of value" (Herzfeld, 2004) on heritage areas, including the increased competition between neighboring localities[19] and the increasing dependence on wider networks of carefully targeted quality consumers. An ethnographic analysis of the heritagization of dairy production thus sketches a complex and dynamic map. In the transition from a subsistence product to a marketable product, determining which foodstuffs become icons of local heritage seems to be the serendipitous outcome of a complex interplay of conflicting political, commercial and institutional agendas. As a result, unequal access to and positioning into a network of contacts, political allegiances, and economic links allows some producers to transform themselves into heritage entrepreneurs, and put some others out of business altogether, while the symbolic construction of alpine food is now entering a stage of iconicization, virtualization and ritualization, parading "saviors" and "testimonials" of the relevant disappearing foodstuffs.

References

Bendix, R. 1997. *In Search of Authenticity: The Formation of Folklore Studies.* Madison: University of Wisconsin Press.

Corti, M. and Ruffoni, C. 2009. *Il formaggio Val del Bitt. La storia, gli uomini, gli alpeggi.* Milano: ERSAF.

De Varine, H. (ed.) 2005 *Les racines du futur.* ASIDC/Les Editions du Papyrus.

Desvallées, A. 1992. *Vagues: une anthologie de la nouvelle muséologie. Textes choisis et presentés par André Desvallées.* Mâcon: Éditions W-MNES, 2 vols.

Di Giovine, M. 2008. *Heritage-scape. UNESCO, World Heritage, and Tourism.* Lanham: Lexington Books.

Grasseni, C. 2007. *La reinvenzione del cibo.* Verona, Qui Edit. Revised and enlarged edition in preparation as *The Reinvention of Cheese.* Oxford: Berghahn.

Grasseni, C. 2011 "Re-inventing food: Alpine cheese in the age of global heritage," *Anthropology of Food*, vol. 8, http://aof.revues.org/index6819.html.

Grasseni, C. 2012a. "Resisting cheese. Boundaries, Conflict and Distinction at the foot of the Alps," *Food, Culture and Society*, 15/1: 23–9.

Grasseni, C. 2012b. Developing cheese at the foot of the Alps, in *Re-imagining Marginalized Foods*, edited by E. Finnis. Tucson: The University of Arizona Press, pp. 133–55.

Grasseni, C. 2012c. Reinventing food: the ethics of developing local foods, in *Ethical Consumption*, edited by J. Carrier and P. Luetchford. Oxford: Berghahn, pp. 198–216.

Grasseni, C. 2012d "Enjeux et jeux territoriaux autour de la patrimonialisation de l'alimentation dans les vallées alpestres italiennes," in D. Crozat, L.S. Fournier,

---

19 As Leitch (2003) already noticed in the case of the *Presidium* of the Lardo di Colonnata.

C. Bernié-Boissard and C. Chastagner (eds), *Patrimoine et valorisation des territoires*. Paris: L'Harmattan, pp. 77–90.

Grasseni, C. 2013. "Produits typiques alpins, écomusées et marketing territorial comme stratégies de distinction" in S. Magagnoli et al. (ed.), *La typicité dans l'histoire. Tradition, innovation et terroir/Typicality in History. Tradition, Innovation, and Terroir.* Bruxelles: Peter Lang, pp. 187–209.

Hemme, D., Tauschek, M. & Bendix, R. (eds.) 2007. *Prädikat 'Heritage'. Wertschöpfungen aus kulturellen Ressourcen.* Göttingen: LIT Verlag.

Herzfeld, M., 2004, *The Body Impolitic: Artisans and Artifice in the Global Hierarchy of Value.* Chicago: Chicago University Press.

Hobsbawm, E. and Ranger, T. 1983. *The Invention of Tradition.* Cambridge: Cambridge University Press.

Jacini, S. 1883. *Frammenti dell'Inchiesta Agraria*, 2nd Edition. Roma: Forzani e C., Tipografia del Senato.

Jones, S. 2005. Transhumance Re-Examined. *Journal of the Royal Anthropological Institute*, 11 (2), 357–9.

Leitch, A. 2003. "Slow Food and the Politics of Pork Fat: Italian Food and European Identity," *Ethnos* 68(4): 437–62.

Lombardi Satriani, L.M. 1973. *Folklore e profitto. Tecniche di distruzione di una cultura.* Bologna: Guaraldi.

McNetting, R. 1981. *Balancing on an Alp. Ecological Change and Continuity in a Swiss Mountain Community.* Cambridge: Cambridge University Press.

Paxson, H. 2008. "Post-Pasteurian Cultures: The Microbiopolitics of Raw-Milk Cheese in the United States." *Cultural Anthropology* 23(1): 15–47.

Rivière, G.-H., 1992. L'Écomusée, un modèle evolutif (1971–1980), in *Vagues: une anthologie de la nouvelle muséologie.* Edited by A. Desvallées. Mâcon: Éditions W-MNES, vol. I.

Stefanelli, G. 1998. *Formaggi della Valle Brembana.* Comunità Montana Valle Brembana. Bergamo: Corponove Editrice.

Chapter 4

Edible Authenticities:
Heirloom Vegetables and Culinary Heritage
in Kyoto, Japan

Greg de St. Maurice

Introduction

No place in Japan is associated with cultural heritage more than Kyoto. The prefecture's 17 UNESCO World Heritage sites draw nearly 50 million visitors annually (Kyoto City Finance Bureau 2008). Approximately one fifth of Japan's National Treasures and Important Cultural Properties are found in Kyoto, as are the headquarters of Japan's main Buddhist traditions, arts and crafts associations, and its most famous geisha quarters (Brumann 2009). Kyoto City's Gion festival float parade, televised nationally every July, has been inscribed on UNESCO's list of Intangible Cultural Heritage. For such reasons, Kyoto is known as the Japanese heart/mind's hometown (*Nihon no kokoro no furusato*) (2009).

Kyoto also plays an important role in terms of the nation's culinary heritage. Traditional Japanese cuisine is said to have its origins in the kitchens of ancient Kyoto (Rath 2010). Contemporary Kyoto's diverse cuisines, including the stylized haute cuisine *kaiseki,* vegetarian temple cuisine, and a style of home cooking called *obanzai ryōri,* continue to be sought after and savored by the nation (Hosking 1996). A key component of these cuisines has been a set of heirloom vegetables known as *kyōyasai*, or Kyoto vegetables.

In this chapter, I examine co-existing claims to authenticity for different groups of *kyōyasai*, the criteria on which these articulations of authenticity are based and evaluated, and what this reveals about culinary heritage in Kyoto and beyond. Based on participant observation in Kyoto as well as informal and semi-structured interviews with farmers, chefs, local government officials, food experts, and consumers, and an analysis of texts marketing Kyoto vegetables, I argue that actors in Kyoto's food industry articulate authenticity in terms of content, process, origin, and historical continuity. My approach, then, does not view authenticity as an essence that is to be discovered, but rather as a quality that is socially ascribed (Handler and Linnekin 1984, Kirshenblatt-Gimblett 1995). Moreover, my analysis shows that authenticity is not arbitrarily ascribed, but is the product of a discourse that touches upon specific socially relevant criteria.

As a collective, *kyōyasai* possess an aura of elegance and mystique linked to their origins in the Heian era (794–1185 CE), when Kyoto was the imperial capital and what is now perceived as "traditional" Japanese culture began to flourish. After World War II, as the global circulation of food and agriculture became more pronounced, standardized vegetable varieties—particularly inexpensive foreign imports—became increasingly common and several of Kyoto's heirloom varieties became extinct. In response to this, and what was perceived as a local agricultural industry under assault, Kyoto Prefecture officially defined "traditional Kyoto vegetables" or *Kyō no dentō yasai*. This term is restricted to varieties that have a local history of being cultivated before the Meiji era (1868–1912 CE).[1] Kyoto's prefectural government went on to assist with the creation of the Kyo Brand for vegetables. Kyoto City, in an attempt to assist those farmers within the city who do not participate in the Kyo Brand, has instituted the Kyoto Seasonal Vegetable Program (*kyō no kodawari shun yasai puroguramu*), a project that supports farmers who grow varieties that range from the *manganji* sweet pepper to broccoli and tomatoes. Because it aims to aid all city farmers, whether their crops are traditional vegetables or not, most of the city farmers that I have interviewed take advantage of the Kyoto Seasonal Vegetable Program in one way or another. At an even more micro level, brands such as the Kamigamo Heirloom Vegetable Research Group's brand or the Mozume brand—though limited to only one or two heirloom varieties— stress provenance from particular neighborhoods.[2]

Origin

Although the label "Kyoto vegetable" appears to set up place of origin as the defining trait of this group of vegetables, the term "Kyoto" (and the abbreviated "Kyo") may be used to signify a number of different, overlapping places. "Kyoto" means capital and its use in branding and marketing nostalgically references the imperial capital of Heian times. Recent surveys have determined that Kyoto has the second strongest place-based brand out of the 47 Japanese prefectures (after Hokkaido) (Brand Research Institute 2010, Nikkei Research 2011).

For chefs and food experts, origin also channels the notion of "taste of place" or "*terroir.*" In addition to agricultural techniques said to have been developed in Kyoto, temperature variations and the quality of local soil and water are identified as key factors that have made Kyoto vegetables distinct and delicious.

---

1   There are 37 varieties of *Kyō no dentō yasai* and 3 more that have subsidiary status.
2   Although other areas of Kyoto Prefecture produce bamboo shoots, those from the Mozume and Tsukahara neighborhoods have garnered a reputation for being of especially high quality. This has enabled them to establish place-based brands for bamboo shoots. The Kamigamo Heirloom Vegetable Research Group has accomplished something similar for the *kamo* eggplant and *suguki* (a variety of turnip usually pickled) grown in the Kamigamo area of Kyoto City.

Environmental dangers, including global warming, are thus a worry for local farmers. The 2011 earthquake, tsunami, and Fukushima Daiichi Nuclear Power Plant disaster have not had a tremendous impact on Kyoto's agricultural industry. I am told that consumers in the Tokyo area temporarily stopped consuming vegetables—including those from Kyoto—after the nuclear power plant destabilization, but aside from produce grown in the Tohoku region—and Fukushima Prefecture in particular—vegetable sales have returned to what they were prior to the catastrophe. If anything, at present Kyoto's produce seems to be even more valued as safe and trustworthy because Kyoto has retained its image of quality and traditionality.

To be sold under the Kyo Brand, vegetables have to be grown within the present day boundaries of Kyoto Prefecture. In order to be eligible to be a Kyo Brand vegetable, a variety must be "Kyotoish" or else must help to expand the market for Brand vegetables (Kyō no furusato sanpin kyōkai 2011). The criteria of "Kyotoishness" refers to Kyoto's aura of mystique and elegance as the ancient imperial capital, but is a quality that is ascribed and perceived in the present rather than an inherent characteristic that can be objectively measured. The brand thus elides the difference between the ancient imperial capital and the much larger entity of contemporary Kyoto Prefecture, capitalizing upon the nostalgic appeal of the former. The primary differentiation that results is between those vegetables grown inside Kyoto Prefecture and those grown outside. The Kyoto Hometown Product Association (Kyō no Furusato Sanpin Kyōkai) actively promotes awareness of the prefectural origins of varieties like the *kintoki* carrot, *mizuna*, and the *manganji* sweet pepper (and more recently, varieties of fruit, legumes, and seafood). The prefectural government publishes recipes and information on the nutritional value of different varieties in pamphlets and on the Internet. The prefectural newsletter, delivered monthly to all households, always includes a recipe in which the main ingredient or the secret ingredient is a *kyōyasai*. During seasonal campaigns, samples of dishes made with Kyo Brand vegetables are given out in local supermarkets and Tokyo's upscale department stores.

On its part, Kyoto City assigns origin to specific neighborhoods of Kyoto City and supports farmers from those neighborhoods who cultivate designated vegetable varieties. Many of these vegetables have names derived from the place they are purported to have first been cultivated. The leafy green *mibuna*, for instance, is said to have been grown for hundreds of years in the area around Mibu temple (Takashima 2003). For Kyoto City, the next best alternative to assigning origin to a neighborhood is designating the city as a whole as the origin of *kyōyasai*. This is why the *shōgoin* cucumber may be seen as authentic when the only farmer who grows it does so in the Kamigamo area and not the Shōgoin neighborhood where the variety was once grown. Kyoto City assists smallholders participating in the *shun yasai* program mainly by providing assistance with marketing and distribution. Through the program, Kyoto City has created a logo for produce grown by participating farmers, established stands for selling local produce in places like subway stations, and disseminated information to consumers about places they can purchase locally-grown vegetables.

The Kamigamo Heirloom Vegetable Research Organization, meanwhile, is a group of about 20 farmers who take turns growing and harvesting seeds for the *kamo* eggplant and a turnip variety called *suguki* that is traditionally pickled. One member, Y-san, explained that the *kamo* eggplant they grow is "the real of the real" because it comes from the right neighborhood (the one with the longest continuous history of growing it), as do the seeds and the farmers themselves. The group does not promote itself as the prefectural and city brands do, but its members are active locally. Using old and new techniques—from word of mouth to websites—they explain the crucial role that the Kamigamo area has played in local food culture, particularly as the point of origin for the *kamo* eggplant and *suguki* (Ikeda 2011). The neighborhoods of Tsukahara and Mozume, meanwhile, claim that one reason for the superiority of their *takenoko* (bamboo shoots) is the earth itself: clay-laden and good for growing the light-colored shoots that fetch the highest prices at the Kyoto Central Wholesale Market's spring auctions. Bamboo shoots from both neighborhoods are said to be distinct—Mozume's yolk-colored, Tsukahara's lemon yellow, the result of soil variations. To maintain the strength of its brand, Mozume farmers only sell superior specimens through their brand; lower quality shoots are not packed in Mozume boxes or sent to the freshly harvested *takenoko* shoot auctions.

According to a recent poll, one of the most appealing characteristics of *kyōyasai* for consumers in Kyoto Prefecture is the fact that they are locally produced, a possible indication of the effectiveness of Kyoto's marketing and educational efforts (Kyoto Prefecture Agricultural Research Institute 2007). Local chefs also discriminate according to place of origin. A chef at a renowned local restaurant stated that authenticity mattered to his boss and his customers and that a "real *kyōyasai*" was grown in the "right place" from seeds "from long ago." In fact, the popularity that Kyoto vegetables enjoy today is in large part due to the efforts of local chefs to revive them in the late 1980s and early 1990s.

Content

Seeds hold the genetic material that supposedly carries the "essence" of the varieties themselves. But seeds alone cannot ensure authenticity based on content. The story of the *momoyama* daikon illustrates this. Though farmers stopped growing this local variety of Japanese radish decades ago, researchers at the Kyoto Prefectural Agricultural Research Institute continued to cultivate it in order to ensure that its genetic material would not be lost. One city farmer whose family had never grown it became interested in growing it and was able to get seeds from the Institute. When he presented his radishes at one of the local agricultural fairs, however, he was criticized because his radishes looked unlike the "prototypical" *momoyama* radish. He then received assistance from another farmer who knew what the radish variety had looked like in the past. For several years they planted these radishes, picking out only those that were close to the ideal until they were able to harvest an entire crop of radishes that resembled "authentic" *momoyama* radishes.

Farmers understand that over time a vegetable variety will change and they take precautions to minimize the degree to which this happens. Many farmers in Kyoto City's Kamigamo area harvest seeds from their own superior specimens. But cross-pollination, changes in local climate, and changes in consumer tastes have resulted in vegetables that deviate from the "ideal" form. The *karami* daikon, for instance, does not possess quite the intensely sharp taste once characteristic of this variety and for which it was named.[3]

Local agricultural fairs provide an opportunity for circulating knowledge of authentic appearance and evaluating local produce accordingly. *Kyōyasai* are held to particularly rigid standards. At one fair I attended, a farmer won a prize because his *takagamine* peppers were closer to the ideal than those of his peers, whose specimens appeared to have mixed with the *manganji* pepper.

We see, then, that appearance—and especially shape—is a key criteria for evaluating the authenticity of *kyōyasai* based on content. When vegetable varieties are selected for the Kyo Brand, they are judged according to the degree to which they reflect Kyoto's "mysteriousness" (Kyō no furusato sanpin kyōkai 2011). The *shishigatani* squash passed this test with flying colors though farmers and agricultural experts widely admit that it is far from tasty: the scaly brown hourglass-shaped squash has appeal because of its history and its distinct appearance. Some restaurants place the squash alongside other *kyōyasai* outside their storefronts, signifying their use of seasonal produce and tapping into Kyoto's age-old aura of mystique to attract customers, though they do not feature the vegetable anywhere on their menus.

Taste is relied upon to a lesser degree in determining authenticity. The *kamo* eggplant, for instance, is touted as having soft skin and firm flesh. One chef told me that she discerns an applelike flavor. On the whole, however, those I interviewed—even experts—were inconsistent as to what specific varieties of *kyōyasai* taste like and why they are so delicious. Perhaps this is due to the subjective nature of taste experiences, as well as a lack of a ready vocabulary for differentiating between the taste of similar varieties. It also indicates that Kyoto's food industry seems to have been less successful in circulating knowledge of taste than it has in articulating the authenticity of local produce. If anything, interviewees seemed to think that, save for exceptions like the *shishigatani* squash, authentic *kyōyasai* would inevitably be more delicious than their counterparts from other prefectures.

---

3   Conversation with Professor Nakamura Yasushi of Kyoto Prefectural University on August 25, 2007.

Process

Process is another criterion according to which authenticity may be claimed or ascribed. Kyoto-grown bamboo shoots illustrate this well. Though Kyoto's soil is said to be critical in setting Kyoto's bamboo shoots apart from those grown in other parts of Japan, the farming techniques that Kyoto's Nishiyama area farmers (including those in Mozume and Tsukahara) use are considered to be another important factor. Kyoto's farmers engage in *tsuchi-zukuri*, soil-making, and "grow" their bamboo using time and labor intensive techniques, in contrast to farmers in other areas who "simply" harvest the bamboo shoots that grow in their bamboo groves. In spite of the value placed on a specific way of doing things, however, process is acknowledged to have changed in some ways that do not diminish a vegetable's authenticity. Certain farming innovations are viewed as positive. For *takenoko*, for instance, Nishiyama farmers now use machines to cut the tops of adult plants in summer, while *kamo* eggplant farmers use grafting techniques, greenhouses, and pollinating bees. Not all changes in cultivation techniques are deemed acceptable for the production of authentic produce, however. Using electricity to stimulate the growth of bamboo shoots, as it is rumored farmers in other areas of Japan do, is seen as cheating that may improve the size or number of shoots but results in inferior texture and flavor.

Patterns are emphasized here, differences between contemporary farmers and their foreparents often de-emphasized or glossed over. Even those farmers whose families have cultivated these vegetable varieties on the same land for centuries, who harvest and plant their own seeds, and who seek to reproduce vegetables that have an appearance close to the "ideal," are without doubt modern farmers. Indeed, the very appeal and designation of *kyōyasai* as "traditional" is a modern (or postmodern) phenomenon. Only in recent times have what were once tasty but "ordinary" local varieties become rare enough that a nostalgic eater may even consider them "exotic."

K-san is one farmer I spoke to who might be called a "traditional" farmer. He grows rare vegetable varieties not because there is a financial incentive to do so, but in order to keep them in existence. He harvests his own seeds, even for the *kamo* eggplant. And yet he too employs modern farming methods. Like most farmers, he grafts the *kamo* eggplant onto a sturdier variety. He limits his use of chemical fertilizers and pesticides and says, "My farming is not 100% organic. That would be impossible."

Continuity

Running down the middle of a Kyoto Brand poster featuring an enlarged photograph of Kyoto vegetables is a list of "past customers": historical figures of national importance like Heian poet Ono no Komachi and nineteenth-cenury politician Sakamoto Ryōma. The text ends with an ellipsis, implying that those who eat Kyoto vegetables today may join this elite list. This sense of continuity is a key ingredient to the authenticity of *kyōyasai*. Rather than stand alone, it is linked to origin, content, and process; the further back in time the continuity of one of these criteria may be traced, the more authentic it is said to be.

Continuity is very meaningful to small-scale farmers whose ancestors grew some of the same varieties on the same land. Several farmers I spoke to grow varieties of cucumber, eggplant, and beans that *at most* one other farmer grows. They choose to continue cultivating these varieties despite the marginal profits involved because these vegetables might go extinct otherwise. Thus many of these farmers have additional sources of income, such as real estate. It is also common for women farmers to take up the distribution routes their foremothers used to sell their produce to regular customers.

What constitutes continuity, however, is defined in the present. This is evident in how the criterion for authenticity that I have labeled "process" is evaluated. Farming practices have changed with the availability of chemical fertilizers and insecticides as well as novel organic techniques. Similarly, traditional vegetable varieties are used in innovative ways: consider *kamo* eggplant jam; *horikawa* burdock root stuffed with wild boar; and the dish that has truly become trendy, *mizuna* salad. At the level of content, experts understand that the varieties are changing visibly and invisibly, in ways that are desirable—even intentional—and in ways that are undesirable or uncontrollable. Indeed, when I visited the Kyoto Prefecture Agricultural Research Institute in 2011 they were setting out to grow *manganji* peppers that are—as advertised and as requested by consumers—more consistently "not hot." Even Kyoto itself—the municipal and prefectural entities; understandings of the ancient capital during various periods of Japanese history; and the ideas, emotions, and fantasies associated with Kyoto—is crafted and re-crafted in the present.

Continuity is not always wholly appealing either. Local food experts argue that changes in consumer preferences over the last half-century have resulted in Kyoto vegetables that are less flavorful than they used to be. Similarly, the women who give out samples of Kyoto vegetable dishes in Tokyo's upscale department stores may proclaim, "These are fine specimens of Kyoto's traditional summer vegetables!" or "Introducing the taste of Kyoto!" but their recipes are adjusted to please the palates of Tokyo residents, who are accustomed to more strongly flavored home-cooking.

Conclusion

These various articulations of authenticity have been economically successful. When it began in 1989 the Kyo Brand sold 38 million yen worth of produce. Sales peaked in 2003 at more than 1.5 billion yen (Kyo no Furusato Sanpin Kyokai 2010). On the whole, vegetables from Kyoto fetch much higher prices than similar varieties from other prefectures. *Mizuna*, which makes up about half of the brand's sales, sells for almost twice the price of that from Ibaraki prefecture, which produces 80 percent of the *mizuna* available at the Tokyo Central Wholesale Market.[4] Although it is difficult to prove causation, local agricultural officials believe that the success of the Kyo Brand has improved the image and the market price for local agricultural products more generally. This is seen as one reason that Kyoto's gross product for fruits and vegetables has suffered less than other prefectures in the Kinki region as Japan's borders have been opened to cheap agricultural imports by a dozen or so free trade agreements (2010).

In the summer of 2009 I saw four cartons of sweet peppers at a wholesale vegetable market in Tokyo. Those from Yamagata Prefecture were selling for 600 yen, those from Kumamoto 800. Meanwhile, a case of slightly misshapen Kyoto *manganji* peppers cost 1,500 yen, and a better-looking bunch 3,000 yen. The store's owner explained that in Japanese cuisine the highest grade vegetables are Kyoto's. "These peppers are totally delicious. The sweetness really comes out. They're not at all like bell peppers," he told me. The price difference he attributed to the strength of the Kyo Brand and its effective advertising.

Consumers may be willing to pay more for authenticity, but authenticity is not equally valued for all products or all aspects of social life. Authenticity is made to matter for issues that are a source of social anxiety. Examining articulations of authenticity reveals political and economic tensions as well as identity crises and cultural friction. Take Northern Italy, for example. In the context of anxiety about the effects of the globalization of food and agriculture (like Japan) and large-scale immigration (unlike Japan), the idea of an "authentic" Italian cuisine is one that resonates. Groups such as Lega Nord and the Slow Food Movement have articulated disparate visions of what such a cuisine should be (McKinley 2010).

Japan currently imports 60 per cent of its food (on a calorie basis), more than any other OECD country (Ministry of Agriculture, Forestry and Fisheries 2010). With the national media pouncing on food scares and largely blaming foreign countries, particularly China, for problems in which domestic actors are also implicated, place of origin indexes a food's trustworthiness, with implications for households and the nation (Rosenberger 2009). Against this backdrop, the attribute and label "domestically produced" adds value to edible goods. Because of Kyoto's position as the "hometown of the Japanese heart/mind," provenance from Kyoto enhances the value of commodities that are convincingly "Kyotoish" all the more. Indeed, claims that the "real" Kyoto vegetables are those grown in Kyoto Prefecture—or more specific areas within the prefecture—emerged in the late 1980s at a time when certain heirloom varieties were rare and had slipped out of the public consciousness while others were being grown in other parts of Japan. By the early 1990s, when the first *kamo* eggplants went on sale in Tokyo, they sold for over $7 USD a piece (Yomiuri Shinbun 1991). Some consumers were willing

---

4   The following website was used to obtain data on *mizuna* sales in Tokyo: http://www.shijou.metro.tokyo.jp.

to pay such a price for something that was culturally meaningful, a sign of luxury, and familiarly exotic.

The success of the Kyo Brand has accelerated the creation of other prefectural brands for produce. Kyoto's farmers and local officials view this as a positive development. Other places also value their agricultural and culinary heritage, and Kyoto's residents appreciate that. Moreover, it is clear that Kyoto is open to food from other places. Japan's self-sufficiency rate for food may hover around 40 percent, but Kyoto Prefecture's has hit a low of 13 percent (Ministry of Agriculture, Forestry and Fisheries 2010). Cooks, be they chefs in traditional style restaurants or the homemakers who create Kyoto's daily dishes, incorporate ingredients and cooking techniques from outside. Hence local Italian restaurant Azakura's *shōgoin* turnip stuffed with parmesan cheese, the Prefectural newsletter's recipe for "spaghetti, tuna, and *mizuna* salad," and a seasonal Kyoto vegetable salad tossed in Dijon mustard based dressing at a workshop on local food culture.

The popularity of *kyōyasai* has prompted farmers in other prefectures and even other countries to grow these varieties or similar ones. But no matter their appearance or even how delicious they may taste, *kyōyasai* varieties grown outside of Kyoto do not possess the authentic taste of Kyoto, redolent with history and nostalgia. It is this authenticity—marketed via posters that proclaim the history of *kyōyasai*, food samples offered in Tokyo's upscale supermarkets, and the promotion of culinary tourism in Kyoto—that has turned *kyōyasai* into value-added craft foods. As one farmer succinctly put it, "If you grow a Kyoto vegetable in California, it's not a Kyoto vegetable anymore."

References

Brand Research Institute, Inc. 2010. *"Chiiki Burando Chōsa 2010" Chōsa Kekka: Motto mo Miryokutekina Shikuchōson ni Sapporoshi ga Kaerizaki! Todōfuken de ha Hokkaido ga 2 Nen Renzoku* [Online: Brand Research Institute]. Available at: tiiki.jp/corp_new/pressrelease/2010/20100908.html (Accessed 24 April 2012).

Brumann, C. 2009. Outside the glass case: The social life of urban heritage in Kyoto. *American Ethnologist*, 36(2): 276–99.

City of Kyoto Investor Presentation. 2008. Kyoto City Finance Bureau, Finance Division, Budget Section.

Cwiertka, K. 2006. *Modern Japanese Cuisine: Food, Power, and National Identity.* London: Reaktion Books.

FY2009 Annual Report on Food, Agriculture and Rural Areas in Japan Summary. 2010. Ministry of Agriculture, Forestry and Fisheries.

Handler, R. and Linnekin, J. 1984. Tradition, genuine or spurious. *The Journal of American Folklore* 97(385): 273–90.

Hosking, R. 1996. *A Dictionary of Japanese Food : Ingredients and Culture*, 1st edition. Rutland, VT: Charles E. Tuttle Co.

Ikeda, T. *Kyō No Tsukemonotachi* [Online]. Available at: web.kyoto-inet.or.jp/org/ suguki/top.html (Accessed 29 November 2011).

Kirshenblatt-Gimblett, B. 1995. Theorizing heritage. *Ethnomusicology* 39(3): 367–80.

Kyō no Burando Sanpin Burando Suishin Jigyō 20 nen no Ayumi. 2010. Kyō no Furusato Sanpin Kyōkai

Kyō no Furusato Sanpin Kyōkai. 2011. *Kyō no Burando Sanpin (Kyō Ma-ku) tte?* [Online: Kyō no Furusato Sanpin Kyōkai]. Available at: kyo-furusato.jp/mark. html (Accessed 26 November 2011).

Kyōtofu Nōgyō Sōgō Kenkyūjo Shiken Kenkyū Seiseki Hōkokukai Hōkoku Yōshi. 2007. Kyoto Prefecture Agricultural Research Institute.

McKinley, L. 2010. *YES TO POLENTA, NO TO COUSCOUS!: Constructed identities and contested boundaries between local and global in Northern Italy's gastronomic landscape*. B.A. honors thesis, School of International Studies, University of Washington.

Nikkei Research. 2011. *Chiiki Burando Senryaku Sa-be-*. [Online: Nikkei Research]. Available at: www.nikkei-r.co.jp/service/branding/area-brand.html (Accessed 28 November 2011).

Rath, E. 2010. *Food and Fantasy in Early Modern Japan*. Berkeley: University of California Press.

Rosenberger, N. 2009. Global food terror in Japan: Media shaping risk perception, the nation, and women. *Ecology of Food and Nutrition*, 48(4), 237–62.

Takashima, S. 2003. *Kyō no Dentō Yasai to Shun Yasai*. Osaka: Tombo.

Yomiuri Shinbun. 1991. Dentō no kyōyasai ga fukkatsushi hajimeta gurume ni kenkō shoku ni. *Yomiuri Shinbun*, Osaka Edition. 1 January, section 8 1.

Chapter 5

The Everyday as Extraordinary: Revitalization, Religion, and the Elevation of *Cucina Casareccia* to Heritage Cuisine in Pietrelcina, Italy

Michael A. Di Giovine

Introduction

Heritage cuisine—a particular form of cuisine which, through its ritualized production and consumption, and ties to the unique milieu in which it is found, binds individuals across time and space through discourses of patrimony and inheritance—is quickly becoming a dominant cultural feature in even the most remote and impoverished regions. Like other forms of heritage, heritage cuisine invokes a sense of inherited "tradition" that must be preserved from the transience of time (cf. Freud 1950: 35)—and particularly in the face of modernity and the "schismogenesis" that seems to threaten the integrity of ethnic groups (Bateson 1935, see Di Giovine 2010a). In order to be passed on to future generations, therefore, heritage cuisine often defines itself against new and globally diffuse, mechanized means of production. As Penny van Esterik (2006) has argued in the context of the Lao PDR, since heritage cuisine often draws on conceptions of locality (however vast the "local" is conceived to be) that are appealing to cosmopolitan Westerners, it is often directed to a non-local, often connoisseur crowd—frequently with expectations of lucrative profit. Thus, while certainly we can, and should, view heritage food as a particular commodity form that continues to impact global markets, it transcends economics, transcends the product's material mobility, and transcends even the consumers themselves. It often reaches deep into the local society itself, impacting it in immaterial, yet lasting, ways. It can sometimes even be perceived as a vehicle for revitalizing a social group that believes itself to be on the verge of cultural extinction, at the hands of modernization, acculturation, or schism.

This chapter is based on nearly three years of ethnographic research on socio-cultural revitalization in Pietrelcina—a Southern Italian village whose inhabitants have begun to reformulate their identity based on a valorization of particular elements of their culture they imagine to have flourished during the late 1800s, a period in which the saint and stigmatic, Padre Pio of Pietrelcina, lived in the town (see Di Giovine 2010b, 2012a, 2012b, 2014). While I have shown elsewhere that this revitalization movement is fueled primarily by the "biographical narrative" (Bourdieu 1987) of this saint and the increasing valorization of the town (and therefore its current-day inhabitants) by the some 600,000 religious tourists that pass through it annually (cf. Di Giovine 2010b), like so many other places in Italy, the valorization of its local, non-elite cuisine—or *cucina casareccia*—elevated to the status of heritage food, remains central to this movement.

Anthony Wallace's Revitalization Theory

In 1956, anthropologist Anthony Wallace proposed his model of a revitalization movement, which he defined as a "deliberate, organized, conscious effort by members of a society to construct a more satisfying culture." It is a particular type of "culture change phenomenon" (Wallace 1956: 265, 267), brought about by cultural contact and transformation at the hands of an outside society, which is fueled by the society's desire to restore ideal cultural values. In this paradigm, society is conceived as a non-reified Durkheimian organism (Wallace 1956: 265–6; cf. Durkheim 1995: 216–24)—one that can be thrown out of equilibrium during periods of anomie. Unlike Durkheim, however, Wallace's societal individuals consciously entertain both a mental image of their society and culture, as well as their own bodies and behavioral regularities—which Wallace defines as a *mazeway*, a mental image of the individual's role in the biological, environmental, social, cultural, historical and supernatural world (2003: 12). When a significant number of societal members feel that their mazeways are in disequilibrium—that is, that their mental image no longer conforms to their lived experiences—they may attempt collectively what Wallace calls "mazeway reformulation," a rectification of a mazeway's inability to help an individual find the appropriate path within a changing environment. Unlike other forms of culture change, the solution posited by a revitalization movement makes explicit recourse to the past when positing the way forward for the future. Based on the society's cosmology, it imagines a set of practices, traditions and values from an idealized point in its history, posits that the contemporary society had abandoned them after contact with outsiders, and proposes a way to reclaim them. If Hobsbown and Ranger defined the "invention of tradition" as practices that "seek to inculcate certain values and norms of behavior by repetition, *which automatically implies continuity with the past*" (2003: 1, emphasis added), then the content of a revitalization movement can be considered a *reinvention* of tradition—a discontinuous break with the immediate past in favor of earlier values or processes that are considered more authentic, but have been deemed lost or obscured. While the society may believe it is simply returning to an unadulterated state, it really creates a radically new worldview that melds imagined past values with contemporary ones.

Revitalization movements also operate with an understanding of society that structurally resembles a seasonal, agrarian life cycle: one's culture, like one's food, can be germinated, flourish, grow sick, and become reborn. Likewise, as Di Giovine and Brulotte mention in the introduction to this volume, a single foodstuff also undergoes multiple revitalizations as it travels through a foodway: from agricultural product to ingredient to dish to cuisine, the original material continually takes on new lives, new significances, and new roles as it moves through different social contexts, impacts, and is impacted by, other actors. As this chapter will show, Pietrelcinese food and the Pietrelcinesi themselves mirror each other in these multiple, nested processes of revitalization. This revitalization sentiment is conveyed in a Pietrelcinese guidebook, whose authors write, "[Pietrelcina's culture] is rooted in the soil that the Pietrelcinese continually cultivates, [and enjoys] a close relationship between planting and fertility, between harvesting and dying, between offspring and seeding. From seeds to fruit, [his culture] coexisted in a sort of ideal continuum, indissoluble with the fruits of the land" (Comitato Festa 2010: 83).

The Decline and Revitalization of Pietrelcina

Pietrelcina is a 3,000-person town in the province of Benevento, the only land-locked province in the Campania region of Southern Italy. Historically, this has been one of the poorest areas of Italy (Davis 1998), and strongly resistant to change—irrespective of the amount of funding provided for targeted infrastructural improvement. Its society is composed mainly of self-sustaining farmers who either owned or rented small patches of arable land outside of the town. While it resembles surrounding towns in its physical composition, aesthetics, and size, since at least the mid nineteenth century Pietrelcina was regarded in the region as an excellent producer of olives and artichokes. It is also known for its collective devotion to the Virgin Mary, whom Pietrelcinesi call the Madonna della Libera because her wooden effigy is believed to have miraculously "liberated" Pietrelcina and surrounding towns from a deadly cholera outbreak in 1854. Today it is best known as the birthplace of St. Padre Pio of Pietrelcina (1887–1968)—a Capuchin friar whose supernatural abilities, highly publicized stigmata, and poverty alleviation initiatives have led him to become "the world's most revered saint" (Wilkinson 2008), one to whom Italians and Irish pray more frequently than to Jesus, Mary, or other Catholic saints (Bobbio 2006, Keane 2007: 200). But in 1918, at age 30, Pio left Pietrelcina for a remote monastery in San Giovanni Rotondo, 165 km away, never to return again. That year, he received the stigmata, catapulting both him and the town of San Giovanni Rotondo to international prominence. In his 50-year ministry—and continuing after his death in 1968—upwards of six million pilgrims visit San Giovanni Rotondo annually, making it by some counts the second largest and second most visited Catholic shrine in Europe (Anon. 2008, Stanley 1998). San Giovanni Rotondo today includes a sprawling shrine complex complete with a mega-church designed by Renzo Piano that can accommodate upwards of 30,000 pilgrims, and is adorned with contemporary art from internationally renowned architects (see Piano 2004, Oddo 2005).

Until recently, however, Pietrelcina was left off of the pilgrimage itinerary. Locals would travel to see Padre Pio, complaining of their suffering and marginalization. He told them, "I have valorized San Giovanni in life, Pietrelcina I will valorize in death" (Da Prata and Da Ripabottoni 1994: 163). This oft-recited quote is believed to be a prophecy for Pietrelcina's regeneration after Pio's death, as well as a promise by Pio to give Pietrelcinesi special spiritual and material benefit by virtue of their kinship with him (see Di Giovine 2012c); however, the period between 1968 and 1998 only saw increased socio-cultural stress and a notable sense of stagnation. Not only were the physical structures left to decay, but a distinctive "flight from the land" occurred as educated Pietrelcinesi left for Northern Italy and abroad in search of work. According to Italian census reports, when Pio lived in Pietrelcina, the town's population was the highest recorded by census, while today, it is the lowest (ISTAT 2013). Older Pietrelcinesi contend that this decline was the result of a change of agrarian values: they had stopped cultivating the "traditional" artichoke in favor of the "dirty, get-rich-quick" tobacco crop (interview, 31 July 2009), a metonym for the sentiment that the younger generations have "forgotten" or lost touch with their traditional values. In the post-war era, tobacco replaced the artichoke, only to create a crisis in the 1990s when Benevento's tobacco processing plant closed, its operations shifting to Eastern Europe.

Fortunately, it was in this period that the movement to declare Padre Pio a saint—which formally began in 1982 when Pope John Paul II issued the required *nihil obstat*—gained momentum. Canonization procedure requires as complete a biographical portrait of the would-be saint as possible so as to verify his holiness, obedience, and sound mental state (see Delooz 1983, Cunningham 2005). Among other documentation, this involves reconstructing his early life, collecting testimonials from friends, family members, and those close to the subject, and analyzing his early writings—all of which engaged Pietrelcinesi more directly in the process. It also integrated Pietrelcina more robustly into new biographical and hagiographic narratives of the would-be saint; prior to this period, there were extremely few books written on Pietrelcina's role in Padre Pio's life (Di Giovine 2012b), but in this period outside researchers began to devote full chapters to describing Pio's childhood in the town, and local Pietrelcinese intellectuals (primarily lay historians and educators) began self-publishing their own research on Pietrelcina's history and Pio's early life. Not only were biographers and those officially involved in Pio's cause for canonization interested in Pietrelcina, but devotees—foreign and domestic—were as well. As the media would report on happenings within the process, interest grew; international visitors whose knowledge of Pio's biography were less colored by the official narratives promulgated by his shrine in San Giovanni Rotondo—and who were thus more receptive to alternative narratives—began to come in the 1980s, while Italian pilgrims and seasoned devotees—whose imaginaries of Pio were strongly tied to their remembrances of interactions with the friar or visits to his tomb in San Giovanni Rotondo—only arose a decade later. But by the mid 1990s, visitation to Pietrelcina reached a fever pitch. "At that time it was so packed you couldn't breathe," said one Irish guide who had been leading tours since the 1970s (Di Giovine 2010b: 279).

The turning point occurred in 1999, the year of Pio's beatification—the last step before canonization. In that year, guides recall, tourism to Pietrelcina reached its frenzy, and a television crew came to the village to film a documentary entitled *Padre Pio: Sanctus* for the Italian national television station, Rai1 (see Damosso 1999). Dedicating a full 16 minutes to Pio's early life and upbringing, this hour-long documentary is groundbreaking because it represents the first to actually use footage from Pietrelcina itself. Marked by slow, languid pans across an intensely verdant countryside surrounding Pietrelcina, the documentary both responded to, and certainly deepened, the imaginaries that had already begun to be constructed of Pietrelcina as an idyllic Italian hill town, not unlike those of the cobblestoned and wine-soaked Tuscany (see Di Giovine 2014). Yet because of Pietrelcina's agricultural downturn, which caused a flight from the land and a lack of funds for the upkeep of the town's cityscape, the documentary's footage revealed visible signs of decline. While Padre Pio's homes and churches had been restored after an earthquake in the 1970s and another the following decade—thanks in part to Italian-Americans and the Connecticut (USA)-based Padre Pio Foundation—the documentary's footage reveals a hodge-podge of rickety buildings, gap-toothed stones peeking from patches of crumbling asphalt, and a central piazza used as a parking lot for old cars. Such sights seemed to have disturbed these idealized images of a bucolic, traditional town, and local leaders commissioned architects from the University of Naples to develop an ambitious plan to sustainably re-create the town (see De Feo 1995). This was certainly economically motivated, but, as I argued elsewhere, it also seemed to have been motivated by the very Italian sentiment of *campanilismo* (town pride), and a sensitivity on the part of Pietrelcinesi concerning how they were portrayed to outsiders (Di Giovine 2010b). The former mayor in charge of the project, Domenico Masone, described his rationale to me:

> It's something so natural and obvious, that when you realize that your clothes are a little older, you buy new clothes to be more presentable. … You bathe, you shave, you prepare yourself for the party. [You do this] not only for the dignity of the place's inhabitants, but also to give a more dignified welcome—to show respect—to those who visit. So that our territory can have dignity, that our hospitality can be comfortable, honest, and acceptable particularly to those who come in a spirit of prayer. (interview, 31 July 2010)

It is true that the endeavor to refashion Pietrelcina focused squarely on the town's urban setting and not directly on its rural countryside; many built structures were completely remade, the houses associated with Pio and his family were restored and repopulated with period furniture, the piazza was enlarged, and all of the streets were pedestrianized and repaved with shining white stones. As one guidebook published by locals contends, "Today, [Pietrelcina] is a town that … is loosing its original agricultural footprint, but trying to play a very important role in religious tourism" (Comitato Festa 2010: 10). But it is important to realize that the two are not mutually exclusive. While touristic efforts primarily involved enormous historic preservation and restoration efforts, they necessarily included re-valorizing and re-packaging the townspeople's intangible resources; and food—primarily *la cucina casareccia*—can be considered its resource *par excellence*.

The Sagra dei Carciofi

For Pietrelcina—whose inhabitants are increasingly considering their ninteenth-century history as a golden age whose values must be re-discovered in order to re-orient themselves within an increasingly globalizing society—the Pietrelcinese artichoke (in reality there are three closely related varieties) is considered the "lost symbol" of local identity:

> The preparation of artichokes in Pietrelcina is one of the town's most ancient traditions, and the artichoke is present in every farmer's garden—as testified by the affirmation that 'stuffed artichokes' was one of St. Padre Pio's favorite dishes. It is a dish of notable nutritional and functional value, and can generate a level of revenue that is competitive with tobacco. It is for this reason that it can be fully integrated into discourses of productive reconversion. (Circelli 2007: 10–11)

This quote is indicative of the artichoke's role in a broader revitalization movement on several levels. First, although research indicates the plant was introduced only in the mid 1800s (Comitato Festa 2010: 102)—several decades before Padre Pio's birth—it is imagined to be deeply rooted in the local culture, one that unites "every farmer" throughout space and time. Second, as has become necessary when re-presenting Pietrelcinese identity today, a link is made to Padre Pio, although I have not found any indication that Pio's family cultivated artichokes. Indeed, Pio's family members were primarily shepherds who, like other Pietrelcinesi, cultivated their own gardens for subsistence purposes in the countryside of Pietrelcina called Piana Romana. Furthermore, there are a number of hagiographic references to Pio being a shepherd as a child (likely because it also fits into Christological metaphors of the "good shepherd" and the *agnus Dei*), and while several biographies include recollections by his childhood friends who would tend their families' sheep together, none mention artichokes. Third, this quote also reveals the tensions inherent in a revitalization movement: between an idealized past (Padre Pio's artichoke) and the present reality (tobacco) rendered non-functional by outside influences (the tobacco market).

As early as 1976, the town council and the Pro-Loco (the town's tourism board) held the first festival celebrating the artichoke, the "Sagra del Carciofo." *Sagra*s are secular food-filled festivals predominantly geared towards celebrating a local product (such as ceramics, flowers, or food), and are typically marketed to a regional crowd. Regarded a "triumph of ancient and authentic flavors of the countryside," the Sagra del Carciofo has also informed the creation of other public festivals, such as the "white night" celebration, *S'adda fa mattina*, a food festival where locals in the Castello district—the neighborhood in the historic center in which Padre Pio lived—set up small *bancarelle*, or food stalls, where they offer inexpensive, local home-cooked meals: pasta, roasted pork, local breads and wine. Importantly, in an attempt to pass on a sense of ownership over the town's "heritage food" to the next generation, the responsibilities for organizing the festival today has been given to the town's youth group, who answer directly to the Mayor's cultural minister. And in 2006, the town obtained additional funding from both the European Union and the Italian government to participate in the regional *Scuole Aperte*—a unique system of "continuing education" in Italy. During the sagra, a master chef holds a class on cooking the artichoke. In conjunction, a pamphlet-cookbook is distributed by the Pro Loco. All of these changes indicate important aspects of a revitalization movement: not only does it make recourse to past processes, but it is decidedly future-oriented; and not only is it intensely local in nature, it is a form of self-representation that is nevertheless oriented outside of the town's boundaries.

Faced with the accomplishments of these touristic endeavors, Pietrelcinesi leaders are pushing for the creation of a local artichoke processing plant, and have already begun the process of gaining protective status for their product:

> If all of these reasons have convinced agricultural producers that they can successfully engage the product on the market, the organization of a short distribution chain and the creation of a local artichoke processing plant, would be a worthy response to a project in the works, which is important for the entire socio-economic fabric of Pietrelcina: that is, to valorize the Pietrelcinese artichoke with an IGT label. (Cercelli 2007: 11)

In short, they are looking to "export" their cultural resource to a broader national and international audience both materially (in the form of "zero-kilometer" products), but also immaterially (in the form of a protected status authentication). In this way, Pietrelcinesi are inserting themselves into a growing movement to designate and protect the particular agricultural products of a place, which they deem to have heightened value. These are government-issued *terroir* designations—ones that ostensibly assure consumers they are enjoying the authentic "taste of a place" (cf. Trubek 2008)—and the earliest and most famous in Italy stem from the wine industry: IGT, or *indicazione di geografica tipica* (typical for the geographic region); DOC, or *denominazione di origine controllata* (controlled designation of origin), which, like the French AOC (*appellation d'origine contrôlée*), require the wine to be produced in a specific well-defined regions, according to procedures that reflect the region's traditional wine-making practices; and DOCG, or *denominzatione di origine controllata e garantita* (controlled and guaranteed designation of origin), which are similar to DOC but whose taste is subject to stringent tests to ensure continuity over time. These protect particular alimentary goods from copycat products (such as Wisconsin-made "Parmigiano-Reggiano" or California "Chianti"). However, we should not simply view these legal designations solely through the lens of commerce. These *terroir* brandings are so culturally situated throughout Italy that they have even infiltrated the Italian lexicon. Today, the term DOC is used as a metonym for "authentic," particularly when talking of a person's identity. One says he's "Pietrelcinese DOC" to mean, "a *real* Pietrelcinese," or "an authentic Pietrelcinese": one who not only has his genetic roots in the place, but manifests the culture's core character traits through his personality.

The *Cicatelli* as Touristic Souvenir and Identity Marker

The Pietrelcinese artichoke has spurred other occasions in which *cucina casareccia* has been elevated to heritage cuisine during the course of this revitalization movement. The most curious is Pietrelcina's "handmade" pasta. Virtually every town or region in Italy identifies itself with a particular form of pasta (or other *"primo"* or first course) that is produced and consumed on a relatively regular basis within the home. For example, Bologna famously has its *tagliatelle* and *tortellini*, Mantova has pumpkin ravioli, Siena has *picci*, and Rome has its *buccatini*. Pietrelcina, like other towns in the province, has *cicatelli*. These are elongated, shell-like pasta that, to me, resemble the mouth of a venus flytrap—though many, such as my family, actually make yarmulke-shaped *orecchietti* (literally ear-shaped pasta, but we call them "pope's hats") but call them *cicatelli*. It is a simple, but extraordinarily time-consuming process to make, since every piece of pasta must be cut into cubes and rolled off of one's fingers; approximately 30–50 pieces are in a typical serving. Like elsewhere in Italy, such time-consuming home-preparation has all but disappeared, as the post-war generations prefer to purchase a wider variety of pre-packaged, often mass-produced pastas such as *Barilla* (which, coincidentally, is thought of as one of the cheapest brands of pasta). At the time of this research, in fact, Barilla had just launched a high-end product line that focused on "authentic" local pastas from the Emilia-Romagna region in which the company is based (see: www.barilla.it/prodotti-barilla/5268/emiliane-barilla).

But *cicatelli* are making a comeback, thanks to a number of entrepreneurial tourist shops owners in the town. Faced with intense competition by both other souvenir shops hawking the same rosaries, statuettes and Padre Pio cigarette lighters (cf. Di Giovine 2009), several stores began to sell "typical foods" from the

Figure 5.1   Making *cicatelli* by hand

Note: These are actually *orecchietti* but this cook calls them *cicatelli.*

Figure 5.2   Nonna Pina's *cicatelli*

Note: Tourists with bags of pasta, which are advertised as "artisanal." Regardless of whether they are homemade or not, at nine packs for EUR 5, they are popular "edible souvenirs" among Italian tourists.

town. By far the most successful are the packages of these seemingly handmade pastas, passed off as a form of heritage food. Strictly speaking, however, they are not; they are actually machine-made, delivered every few days by factory trucks.

But the re-presentational process by which these *cicatelli* are converted into to heritage cuisine is important: they are sold in bags with handwritten labels with such names as "Nonna Pina's" ("Grandma Pina's"), and stating that it has been produced in Pietrelcina; the selection of this particular name may have also been a strategic one, as there is a popular contemporary children's song called "Le taglietelle di Nonna Pina" ("Grandma Pina's tagliatelle"). The pastas themselves are irregular in shape, one seemingly different from the other, and do not have a finished or processed look. And while each store has a slightly different offer, the prices are extraordinarily inexpensive—another marker of true home production: eight or nine bags for EUR 5; at about 70 cents per bag, they beat out even the cheap Barilla boxes. These are by far the best-selling "souvenirs" in Pietrelcina among Italian visitors: busloads of Northerners and Southerners alike will be seen carrying between three and four shopping bags filled with the pasta.

Elements of *Cucina Casareccia* in Special Feasts

Regardless of whether or not these particular *cicatelli* are actually homemade, we must understand that their importance lies not solely with their commercial value, but as another food-based marker of Pietrelcinese identity. Through their juxtaposition in stores with other local heritage foods, they simultaneously gain and give value to the other foods such as salami, cheese, wine and sugar-coated *taralli*—little doughnut-shaped pretzels. *Taralli* are normally *salati*, a savory food that, in the region, is often made with fennel seeds and salt, but on holidays such as Christmas Eve they are often sweetened with chocolate, honey, or pomegranate seeds (Di Giovine 2010a: 202). Informants in Pietrelcina have begun to talk more of their particular form of sweet taralli called *raffiuoli*, which, at the time of Padre Pio, were served during weddings. By their own account, locals contend that the production of these treats had all but died out, but, once again, they were "rediscovered" and inserted into the Pietrelcinese narrative of Padre Pio:

> *Il raffiuolo* is a product that Pietrelcinesi use during moments of celebration, such as a wedding or when a priest is installed like our great monk Padre Pio. (amateur archaeologist, 15 November 2010)

> … on the occasion of Padre Pio's ordination … [there was a] huge festival when he arrived in the town; joy, ebullience, and congratulations met the priest as he timidly entered the town, red-faced … and the people, watched from their windows and balconies as others threw money, rice and 'raffiuoli,' a typical local sweet, as they would do to wish a bridegroom good luck. (Comitato Festa 2010: 95)

Today, weddings in Pietrelcina once again feature these raffiuoli, as one shop-owner, who offered me an unusual version of the typical sugar-coated Jordan almond (*confetti*) from her daughter's wedding the week before, stated.

> ME: Oh, these are really good. You gave them out at the wedding?
> W: Yes, but naturally we also had *raffiuoli*! Do you know what those are? They're a special treat only from Pietrelcina.

I mention this primarily because the *raffiuoli* have become an important example of the dialectical workings of the process of elevating home-cooked food to a heritage cuisine. While the *taralli* sold to tourists as a prized, authentic specialty are decidedly not *raffiuoli*—they are smaller and are created with a different recipe—I believe that they are a sort-of a "secret" stand-in for the local *raffiuoli*; they were a way of re-presenting a traditional delicacy, which could only be enjoyed on special occasions among locals, for quotidian consumption by non-locals. Their prominence prompted others to "re-discover," and indeed elevate, this rather peasant dessert as a highly valued food.

Figure 5.3   **Basket of traditional food at the altar below the effigy of Pietrelcina's patroness, the Madonna della Libera, after its presentation during the Mass on her feast day**

Diffuse once again at local weddings, the "re-discovery" (or "re-invention") of *raffiuoli* is ultimately a local process, experienced for and by locals. They occupy an important place in local culture, despite (or in addition to) the fact that other versions—such as the sugarcoated *taralli*—may be packaged as "heritage cuisine" for more global markets. The presence of food at Pietrelcina's patron saint's feast day is a case in point. Held the first weekend of August, the feast of the Madonna della Libera has, since the mid 1800s, been a harvest feast. In accordance with popular religious practices of Southern Italy, in which devotees often donate valuables for graces they have received or hope to receive through the saint's intercession, the agrarian majority of the citizenry would donate their foodstuff—usually wheat or grain—in mule-drawn carriages, while members of the highest class would donate money for the festival. This process has been modified a bit today. Gone are the often elaborately painted carriages filled with just-harvested wheat. Gone are individual offerings of food. But during the "Mass of Thanksgiving," the leaders of the Festival Committee proceed to the altar with a big basket filled with food. Its contents are telling—they include homemade *capicolo Leuzzo* and salami, olive oil, a bottle of earthy Aglianico wine (which is increasingly being cultivated for export by large, formerly tobacco-producing farms), a mild local cheese called *cacciocavallo*, honey from the countryside, and, of course, a jar of pickled artichokes.

The elevation of these previously everyday foods to forms of heritage cuisine fitting for use as a public sacrificial offering to Pietrelcina's protector-saint tellingly illustrates the revitalization movement's inherently dialectical process. A revitalization movement is not a rediscovery of past traditions—as its members believe—nor is it merely an "invention of tradition;" rather, it is a re-invention of tradition, in which social elements perceived as traditional are re-appropriated, thereby changing their meaning, as Table 5.4 (below), shows.

Figure 5.4   Conversion of everyday food into a form of "heritage cuisine" in Pietrelcina

Cucina casareccia	Heritage cuisine
Produced inside the home for insider consumption	Produced outside the home (factory/shop) for outsider consumption
Socially stratifying marker of poverty	Universal marker of Pietrelcinese identity
Everyday cuisine with little value	Extra-ordinary, valorized as "heritage cuisine"

And so these foods, and their place in Pietrelcinese identity, come full-circle: All of these are manifestations of *cucina casareccia*, but all have become valorized beyond the boundaries of the town as heritage foods, before entering back into the local lexicon, local festivals, local life. In so doing, they both index and construct the cyclical revitalization process of the society that embraces them.

Conclusion

This chapter argued that heritage cuisine can play a significant role in cultural revitalization movements within small-scale societies. The members of Pietrelcina view their society as an organism, and perceive it as sickly and in danger of perishing at the hands of outside influences. These outside influences, they seem to believe, have caused their culture to career "off-track," having lost sight of its moral foundations. Their movement does not revolve solely around food—but neither does it revolve solely around other more obvious elements, such as the figure of Padre Pio, of popular devotion, of pilgrimage and tourism. Indeed, for a revitalization movement to truly be totalizing, to truly be sustainable, it must reach deep into the culture; it must affect the prominent processes and markers of daily life (Di Giovine 2010b). Food is such a marker.

Food has a natural affinity to revitalization movements. On the one hand, revitalization conceptualizes society in agrarian terms: as a raw material, society is germinated, grows, decays, and can either die out or be reborn. Both gain further definition—and sometimes a new life—from how it is elaborated upon during the course of its life. On the other hand, food is, of course, a powerful device for the articulation and negotiation of individual and group identity (Wilk 1999: 244), particularly among communities in various stages of identity crises (Di Giovine 2010a, forthcoming). From preparation to consumption, food serves as a "highly condensed social fact" that embodies claims of "solidarity and community, identity or exclusion, and intimacy or distance" (Appadurai 1986: 498), which, through the action of ingestion, become viscerally embodied in us. Pietrelcinesi use food to create representations and ethnic boundaries that set themselves apart—apart from San Giovanni Rotondo, their chief "competitor" in the cult of Padre Pio; apart from other Italians; and, increasingly, apart from foreigners who visit the town. Pietrelcinesi are also cognizant of the growing immigration and labor problems in Italy and the greater European Union; and North Africans, Eritreans, and even Chinese have set up competing shops in the provincial capital of Benevento and even in neighboring villages. The words of an entrepreneur from Benevento reveal just how heritage food has become the newest and almost "last-ditch" hope for survival amid this dizzyingly globalizing milieu:

> What were we [Italians] previously known for? Italian leather shoes, Italian fashion, cars—now all of these are made more cheaply by the Chinese. We can't compete anymore! ... But what is the one thing that we have that the Chinese can't replicate? Our food. Greco di Tufo—that wine must be made in the volcanic soil around Benevento; if you take the grape and plant it somewhere else it

isn't the same. Our *capicolo* [a form of salami], the olives, the artichokes—to have the same quality, the same taste, they must be made here, in this particular land, with these particular minerals. Our future lies in the traditional products of the land.

Indicative of a revitalization movement, the speaker—like so many others in Pietrelcina—advocates radical changes in the way in which daily life is lived, in which practices can sustain locals, in which values will nurture them, but looks to the past, re-inventing traditions for a way forward. And as this speaker reveals, there must also be a strong identification with the land itself. It is for these reasons such local food is embraced, elevated as heritage cuisine, and re-presented to both those inside the community and outside. And it is in this way, it seems, that the newly valorized marker of Pietrelcinese identity can help revitalize the culture, creating a new, healthier, and more centered society within such a rapidly globalizing world.

Bibliography

Anon. 2008. Padre Pio Set to Beat Lourdes. *Italy Magazine.* www.italymag.co.uk/italy/puglia/padre-pio-set-beat-lourdes (accessed 8 May 2008).

Appadurai, A. 1986. *The Social Life of Things: Commodities in Cultural Perspective.* Cambridge: Cambridge University Press.

Bateson, G. 1935. Culture Contact and Schismogenesis *Man*, Vol. 35 (Dec), 178–83

Bobbio, A. 2006. I più amati dagli italiani: i santi nella storia. *Famiglia Cristiana*, vol. LXXVI, No. 45. 5 November 2006, 66–9.

Bourdieu, P. 1987. The Biographical Illusion. Translated by Yves Winkin and Wendy Leeds-Hurwitz. In *Working Papers and Proceedings of the Center for Psychosocial Studies*. No.14.

Circelli, L. (ed.) 2007. *Progetto "Scuole Aperte": La Lavorazione del Carciofo.* Pietrelcina: L'Istituto Comprensivo "San Pio da Pietrelcina."

Comitato Festa "Maria SS. Della Libera" di Pietrelcina 2010. *Pietrelcina Città Santa. Guida ai Luoghi Natali di San Pio.* Pietrelcina: Comune di Pietrelcina.

Cunningham, L. 2005. *A Brief History of Saints.* 1st (ed.). Malden, MA: Wiley-Blackwell.

Da Prata, L. and A. Da Ripabottoni. 1994. *Beata Te, Pietrelcina.* Pietrelcina (BN): Convento "Padre Pio"—Frati Minori Cappuccini.

Damosso, Paolo. 1999. *Padre Pio: Sanctus.* Rai (documentary film).

Davis, J. 1998. Casting off the 'Southern Problem': Or the Particularities of the South Reconsidered. In Schneider, J. (ed.), *Italy's Southern Question: Orientalism in One Country.* Oxford: Berg, 205–24.

De Feo, Carlo Maria (ed.) 1995. *Pietrelcina: Memoria, tradizione, identità.* Naples: Florio Edizioni Scientifiche.

Delooz, Pierre. 1983. Towards a Sociological Study of Canonized Sainthood. In Stephen Wilson (ed.), *Saints and Their Cults: Studies in Religious Sociology, Folklore and History*, 189–215. Cambridge: Cambridge University Press.

Di Giovine, M. 2009. Re-Presenting St. Padre Pio of Pietrelcina: Contested Ways of Seeing a Contemporary Saint. *Critical Inquiry*. Special edition, "The Curious Category of the Saint." Vol. 35, No. 3, Spring, 481–92

Di Giovine, M. 2010a. *La Vigilia Italo-Americana*: Revitalizing the Italian-American Family through the Christmas Eve "Feast of the Seven Fishes." *Food and Foodways*. Vol. 18, No. 4, December 2010, 181–208.

Di Giovine, M. 2010b. Rethinking Development: Religious Tourism to St. Padre Pio as Material and Cultural Revitalization in Pietrelcina, Italy. *Tourism: An International Interdisciplinary Journal*. Vol. 58, No. 3, November, 271–88.

Di Giovine, M. 2012a. A Tale of Two Cities: Padre Pio and the Reimagining of Pietrelcina and San Giovanni Rotondo. *Textus: English Studies in Italy*. 1, 157–69.

Di Giovine, M. 2012b. Making Saints, (Re-)Making Lives: Pilgrimage and Revitalization in the Land of St. Padre Pio of Pietrelcina. Unpublished Ph.D. thesis, Department of Anthropology, University of Chicago.

Di Giovine, M. 2012c. Passionate Movements: Emotional and Social Dynamics of Padre Pio Pilgrims. In D. Picard and M. Robinson (eds), *Emotion in Motion: Tourism, Affect and Transformation*. Farnham: Ashgate, 117–36.

Di Giovine, M. 2014. The Imaginaire Dialectic and the Refashioning of Pietrelcina. In N. Salazar and N. Graburn, *Tourism Imaginaries: Anthropological Approaches*. Oxford: Berghahn, 147–71.

Di Giovine, M. (forthcoming). Food, Ethnic Imaginaries, and the Ritual (Re-)Construction of Italian-American Identity in the *Vigilia di Natale*. In E. Messina (ed.), *Italian-American Culture and Behavior*. NY: Springer Publications.

Durkheim, E. 1995. *The Elementary Forms of Religious Life* (trans. Karen E. Fields). Free Press.

Freud, S. *The Standard Edition of the Complete Psychological Works of Sigmund Freud. Volume XIV* (trans. J. Strachey). London: The Hogarth Press.

Hobsbowm, E. and T. Ranger. 2003. *The Invention of Tradition*. New York: Canto.

ISTAT 2013. "Demografia in Cifra: 1992–2001." http://demo.istat.it/ricbil/index1.html. Accessed on 5 July 2013.

Keane, C. 2008. *Padre Pio: The Irish Connection*. Edinburgh and London: Mainstream Publishing.

Oddo, M. and Renzo Piano Building Workshop. 2005. *La Chiesa Di Padre Pio a San Giovanni Rotondo [The Church of Padre Pio in San Giovanni Rotondo]*. Milano: F. Motta.

Piano, R. 2004. Padre Pio Pilgrimage Church. http://www.arcspace.com/features/renzo-piano-/padre-pio-pilgrimage-church/.

Stanley, A. 2008. Saint or No, An Old-Time Monk Mesmerizes Italy, *New York Times World*, 24 September 1998. www.nytimes.com/1998/09/24/world/san-giovanni-rotondo-journal-saint-or-no-an-old-time-monk-mesmerizes-italy. html?pagewanted=print&src=pm (accessed 22 December 2010).

Trubek, A. 2008. *The Taste of Place. A Cultural Journey into Terroir.* Berkeley: University of California Press.

Van Esterik, P. 2006. "From Hunger Foods to Heritage Foods: Challenges to Food Localization in Lao PDR." In R. Wilk (ed.), *Fast Food/Slow Food.* Lanham: AltaMira Press.

Wallace, A.F.C. 1956. "Revitalization Movements." *American Anthropologist,* New Series, 58(2), 264–81.

Wallace, A.F.C. 2003. *Revitalizations and Mazeways: Essays on Culture Change, Volume 1.* (ed.) Robert S. Grumet. Lincoln, NE: University of Nebraska Press.

Wilk, Richard. 1999. "Real Belizean Food": Building Local Identity in the Transnational Caribbean. *American Anthropologist.* Vol. 101, No. 2, 244–55.

Wilkinson, T. 2008. Padre Pio Exhumed for a Second Life. *Los Angeles Times*, April 25, 2008. http://articles.latimes.com/2008/apr/25/world/fg-padre25 Accessed 26 August 2008.

Chapter 6

Take the Chicken Out of the Box: Demystifying the Sameness of African American Culinary Heritage in the U.S.

Psyche Williams-Forson

Student: I want to write my paper on African American foodways.
Instructor: That's fine. Any thoughts on what foods you want to consider or the argument you want to make?
Student: Well, I'm thinking about African American food from a Guyanese perspective.
Conversation with student,
"Black Class from the Harlem Renaissance to Hip Hop" (2009)

Frequently, I teach a course titled "Black Class from the Harlem Renaissance to Hip Hop." In it, we explore the complicated designation of "Black" and what it means in the United States to be identified by this socially constructed label. Not surprisingly, our thoughts turn to food habits and traditions as these are among the most contentious and complicated aspects of identity. While the exchange between my student and me, which opens this chapter, seems relatively simple and straightforward it took some time for us to iron out not only her final paper topic but also the ways in which this subject matter helped to complicate our thinking about identity.

Racial politics in the United States tend to view people of the African Diaspora as a monolithic group. To this end, the practices of cultural heritage in which we engage—dances, oral traditions, festive events and even our food cultivation and preparation rituals—are assumed to be the same. This chapter examines some of the foods acquired, prepared, and consumed by "Black" people in the United States to ask what practices are overlooked, who is left out of the discussions, and to what ends when we reduce a diverse set of people to one group of foods? By taking this approach, this chapter argues for a need to complicate our understanding of African American/Black culinary culture and to take a broader look at the study of food culture as heritage practice.

In his now classic essay, "Cultural Identity and Diaspora," Stuart Hall (1990: 222) uses cinema as the frame to argue for problematizing representation and cultural identity. Hall writes,

> Practices of representation always implicate the positions from which we speak or write—the positions of enunciation. .... Identity is not as transparent or

unproblematic as we think. Perhaps instead of thinking of identity as an already accomplished fact, which the new cultural practices then represent, we should think, instead, of identity as a 'production,' which is never complete, always in process, and always constituted within, not outside, representation. This view problematises the very authority and authenticity to which the term, 'cultural identity,' lays claim.

Hall's observation has relevance to this chapter from the standpoint of thinking about the ways in which the foodways among people of African descent are represented and defined by popular culture, and therefore usually limited in scope. Brown and Mussell (1984: 3) assert that Americans use the foodways of the "other" as a "set of convenient ways to categorize ethnic and regional character." At issue here, are the various ways, despite limited perceptions, in which cultural heritages are celebrated in the United States between and among people who define themselves as or are defined as "Black" and/or "African American." Like other countries where a census is used, the United States narrows people of various ethnicities to a single racial category—White, Black or African American, American Indian or Alaska Native, Asian, and Native Hawaiian or Other Pacific Islander.[1] This kind of singularity and distortion in thinking permeates every aspect of our lives from economics to the quotidian act of eating. These taken-for-granted labels of identification function primarily as tools of social and cultural classification, or knowledge soothers, enabling people to sleep at night. However, they can also serve to enable people to coalesce around common experiences and histories. Again, Hall's (1990: 223) arguments have salience here when he suggests that these experiences offer "shared cultural codes," which provide "stable, unchanging and continuous frames of reference and meaning, beneath the shifting divisions and vicissitudes of our actual history." For Black people of the African Diaspora in the U.S., the trope of "Mother Africa" as our central point of origin "is the truth, the essence, of ... the black experience," and it continues to serve as "a very powerful and creative force in emergent forms of representation amongst hitherto marginalised peoples." Often overlooked, but nonetheless present, are the ways in which discussions of foods produced and consumed by African American or Black people in the United States are circumscribed by this frame.[2]

---

1   According to the U.S. Office of Management and Bureau (OMB), race and racial categories included in the census questionnaire "generally reflect a social definition of race recognized in this country and not an attempt to define race biologically, anthropologically. OMB requires five minimum categories: White, Black or African American, American Indian or Alaska Native, Asian, and Native Hawaiian or Other Pacific Islander." http://www.census.gov/population/race/. Accessed January 2, 2012.

2   This study recognizes the problematics of using the terms Black and African American interchangeably. However, the reductionist notion of Blackness based on phonological attributes is at play here. Skin tone and color as well as other obvious physical characteristics have been used to render a person black regardless of their geographical

As a cultural group, with common historical experiences, the collective identities of African Diasporic peoples is centered in part around the struggle to define our place in U.S. society (Lamont and Molnar 2001: 32). But these ethnic identifications and boundaries are shifting and ever evolving rather than fixed and static (Barth 1969). For example, some immigrants from Africa and the Caribbean identify as American Black—both racially and ancestrally. Other times, these same immigrants will resist racial categorization and emphasize ethnic identities that are culturally distinct from American Blacks (Okamoto 2003: 811, cf. Waters 1994). Foodways is a rich site for exploring shifting layers of identity because it is part of a flexible system in which symbols can be replaced, and meanings can change and perform different functions (Kalcik 1984). Foodways practices—procurement and acquisition, preparation, serving, eating, and other traditions/habits—can illustrate not only how groups are sutured together and thus embody a common group identity but also reveal the separations within.

Foodways in the New World

Because most African people entered the New World through the horrific mouth of slavery, it is a natural starting point for considering African American relationships to food and the development of group consciousness. From near starvation to relative abundance, the food practices of early African and Caribbean people in America necessarily influences what we today know as "soul food," Creole soul, Gullah foodways and the like. The foodways of these early generations have obviously morphed and changed as a result of location, relocation, admixing, and acclimation. But it is because of these factors that African American food habits cannot be considered singular in form.

From the seventeenth to the nineteenth centuries the slave trade brought an ever-evolving African American food system. The literature on African American foodways during enslavement is vast and will not be reviewed here except to emphasize that what emerges are two important variables—adaptation and creativity (Singleton 1985, Ferguson 1992, Deetz 1996, McKee 1999). During the Middle Passage, wherein Africans were taken from their homelands, exchanged for goods, and sold into slavery, many captured people experienced starvation, vitamin deficiencies, and even death. Given the trauma of their experience many arrived at their destinations malnourished, if they survived at all. They would have to adjust to their new surroundings including the foods available to them. Entry in the New World economy bore witness to foods being highly regulated, carefully watched, and minimally dispersed. This reality meant new ways of acquiring enough food to survive. The lack of familiar surroundings, ingredients,

origins. This chapter uses the term Black to refer to a collective identity, based on a shared historical experience, that organized around political and social struggles of U.S. society (Hall 1997: 56–7).

and utensils with which to cook and eat also meant that African traditions had to merge with those native to the New World and the colonial settlers occupying their lands. Considerable archeological evidence has been found to indicate that slaves preferred hollow containers like bowls and cups. This, coupled with evidence of small cut animal bones (an indication that meats were of a lower grade), and the knowledge of West African preference for liquid-based meals like soups and stews, indicates a continuation of some African culinary traditions. What emerged over time were creative contributions that emanated from the major foods, herbs, and spices that were available (Singleton 1985, cf. Ferguson 1992, Deetz 1996, Samford 1996, Franklin 2009).

The magnitude of these culinary influences and adaptations reveal themselves when we step away from the myths that all Africans and their descendants ate and continue to eat the same foods. Rather, the reality of different systems of slavery makes this point both farcical and impossible. Slavery was dependent upon environmental settings, which directly contributed to the growth of plantation economies. A lot depended upon the provisions given by the plantation owner but also foods to which slaves had access. Most slaves were allotted a small portion of meal, molasses, and undesirable pork parts. Some planters allowed their slaves to own gardens and to have access to livestock, which provided vegetables and chicken to eat or to sell. But eating also depended on slave's work schedules and weather conditions. On rice and cotton farms of Georgia, for example, a task labor system existed wherein slaves worked particular jobs. Once their tasks were completed slaves sometimes could use the remainder of the day for their own purposes including acquiring their own food. Some enslaved people were given permission to hunt in order to supplement their diets. These activities were few and far between, however, because many plantation owners felt these tasks took time away from work. Meanwhile, some slaves who worked in the houses of their owners may have been afforded treats like tea, coffee, and brown sugar—but these goods were distributed primarily during the holidays (Ball 1858, Singleton 1985, Pulsipher 1990, Yentsch 1995, Walsh 1997, Morgan 1998, Heuman and Walvin 2003).

In short, the experiences of slavery fractured the culinary experiences of African Diasporic peoples, all of who hailed from different ethnic communities in their native countries. In the wake of slavery, however, new enclaves and new ethnic identities were created and with it new struggles and ways of surviving. Through the process of oral transmission, recipes were changed as local ingredients were substituted and infused. Jessica Harris (1989, xvi) and Robert Hall (2007) note spices were used "to disguise spoiled meats and to enhance flavors." More vegetables were added to the diet, as well as cooking techniques including grilling and roasting using cabbage instead of banana leaves. And, different grasses and herbs were used as much for medicinal purposes as they were for enhancing taste variety. This kind of culinary inventiveness is a central aspect of our culinary heritage. From the African imprint left on the foodways of the New World to the creative means of acquiring, producing, and distributing foods, African American foodways has never been as simple as cooking and consuming pork scraps. Rather, for African American women and men, foodways were a major vehicle for the expression of culture and identity (Franklin 2001, 91; Tipton-Martin 2008) due in no small part to the ways in which foods were stretched, augmented, made tasty and sustaining, while filled with ritual and tradition. Politically, this collective identity is one in which Blacks—whether native born or migrant—rarely have been able to escape (Nagel, 1994, Omi and Winant 1994, Waters 1994, Chai 2005: 376–8). Yet, far from homogeneous, this shared history is as much embraced as it is disavowed and food often factors as a contested symbol at the center of these interracial and cultural debates.

Keep on Keepin' on—Soul Food

> [The song] 'Soul Man' was telling a story. Whatever you are doing in your life you are able to get up and keep moving.
> —Sam Moore, singer, Sam & Dave, *Respect Yourself: The Stax Records Story*

Soul Food. This ubiquitous, but problematic (Whitehead 1992, Harris 1995, Paige 1999, Nettles 2007, Opie 2008), label tends to be attributed to foods eaten by African Americans specifically and Black people more generally. The omnipresence of the term may stem from the fact that almost any food—from oatmeal to yogurt—that serves to comfort can be considered "food for the soul."[3] But when the phrase first emerged in the 1960s it was used to describe cuisine eaten by most African Americans and more importantly, it served to signify an identity. As Sam Moore states, the song "Soul Man" told an important story. So too did the phrase "soul food." Both were suggestive of a need to be able to claim a sense of Black pride at time when "Blackness" was constantly questioned and more often threatened. Whether soul referred to food, clothes, music, or other aspects of Black cultural expression it signified that untiring ability to "get up and keep moving." Knowing this, Amiri Baraka (then known as LeRoi Jones), responded that his Black people had soul food to their credit, when a young African American man wrote in *Esquire Magazine* that we had neither a characteristic language nor cuisine. In his collection of essays titled *Home*, Baraka (1968: 121) describes

---

3   A generic search on any of the popular Internet search engines will yield definitions of the term ranging from foods associated with African American vernacular, to articles found in Better Homes and Gardens, to Quaker Oats advertising oatmeal as "soul food"—good for the heart and soul. This leads me to believe that now more than ever Witt's observation that any discussion of soul food needs to be situated within the "context of ongoing intraracial practices of culinary regulation" is particularly salient. And yet, to situate this discussion within those contexts would silence the voices of those who are at the center of the cultural practice described by the array of competing discourses over "soul food."

what we know today as food eaten by many people in the South, and indeed the world: grits, hoppin' John (black-eyed peas and rice), fried fish and chicken, buttermilk biscuits, dumplings, lima beans and corn, string beans, okra, smoky, hot barbecue, everything from the hog—from neck bones to pork chop, sweet tea or lemonade, and sweet potato or other pies and cakes. Though these foods now serve to demarcate a Southern identity (generally speaking) that is "fragile" and becoming "increasing Americanized" (Latshaw 2009: 108) in the 1960s they were representative of a cultural identity directly tied to survival and the battle for liberation and self-dignity.

Food serves as a symbol of a particular kind of group performance and a signal for structuring interactions. Even if African Americans routinely ate the foods we associate with "soul" in their own homes, doing so beyond these confines may have affiliated them with Black Nationalism but surely would have marked them as Southern (Williams-Forson 2006: 80–113). The decision to eat or drink foods associated with "soul" operated "to help individuals develop a sense of selfhood and to confront problems" they were encountering in society despite the stereotypes associated with these foods (Kalcik 1984: 45). Elsewhere, for example, I discuss how chicken, as one food associated with typecasting of African Americans is used to levy assault and at the same time used to mark in-group pride (Williams-Forson 2006). Rather than shying away from those foods with which they had been negatively associated and labeled, some Blacks who ascribed to theories of national pride closed ranks by accepting the stereotypes as a badge of honor denoting their shared oppressions but also their shared resistance to life's inequalities (Williams-Forson 2006, cf. Di Giovine 2010: 182–3). In this way, soul food was used in the social negotiation for power during the 1960s and eating habits were used to make political statements. Fredrik Barth refers to this structuring as "rules that govern inter-ethnic relations" in the maintenance of boundaries. Barth states, "ethnic groups are categories of ascription and identification by the actors themselves, and thus have the characteristic of organizing interaction between people" (1969: 10). And yet, such rules allow for the diverse performance of identities because not everyone participated.

Of course, political reasons were not the only reasons African Americans ate these foods. As Jessica Harris (Chavis 2008) explains by voicing her own ambivalence toward using term "soul food," for many this was just another meal:

> Most African Americans in my experience did not say, 'I'm going to get some soul food' until the sixties. Bottom line, it was dinner. All cultures have food for the soul, and this is just the food for the soul for African Americans ... All soul food is Southern, but not all Southern is soul food, in the sense that most African American food, in the narrow sense, has its genesis in the South. ... There are

dishes in the Southern culinary lexicon that are not soul food at all—Country Captain, for example.[4]

Harris' observation implores us to see this experience of marking/unmarking as a dialectic—part of a complex web of social relations. Doing so, we embrace the heterogeneity of "Blackness" which necessarily contributes to competing discourses about terms like soul food (Witt 1998, 1999: 80). As Doris Witt rightfully notes,

> these discourses span historically far-ranging social and political positions that have been obscured because discussions of 'soul food' tend to rely on simplistic, often insufficiently historicized dichotomies of master versus slave, white versus black, and especially black bourgeoisie versus ghettoite.

While the types of foods African American people eat can vary from region to region several of these foods when eaten in certain combinations can symbolize or mark one as belonging to or having affinity with a particular ethnic group. As a result, one's eating habits often signal their ethnic boundaries. Major deviations from the norms can be read as a rebuff to the community (Kalcik 1984: 47) but it might also be seen as the employment of different styles of culinary engagement by members within the group.

Black People Eat More Than Fried Chicken … Creole Soul

> Gumbo is the soup of New Orleans with soul in it.
> – Wayne Baquet

Though African Americans have migrated—forcibly and voluntarily—throughout history, the first Great Migration of African Americans is said to have taken place from roughly 1910–1930. During this time, approximately 6 million African Americans left the Southern United States moving northward but also westward (to Los Angeles and other Western cities). According to migratory strands, those who left Mississippi and Alabama journeyed to Chicago and other urban spaces. Travelers leaving Louisiana, Arkansas and Texas most likely went to parts of California, Missouri, Wisconsin, and Minnesota. Floridians and Georgians often moved along the eastern seaboard up the coast to the DelMarVa region (Delaware, Maryland, Virginia, and Washington, D.C.), while others continued on up to New England. This brief account serves to emphasize the point that as people moved, they took their foods and their customs with them. As a deeply embedded aspect of one's life, foodways generally are resistant to change (Kalcik 1984: 39;

---

4   Country Captain, a chicken stew flavored with curry, is a dish that likely found its origins in India.

cf. Levenstein 1988, Giddens 1991: 81, Gabaccia 1998). A recent example of this phenomenon took place during Hurricane Katrina. The stories involving how people found shelter go hand and hand with how they located food. Amidst the conversations of destroyed fishing boats and food stamp acquisition is a conversation about missing familiar foods and tastes that remind one of home. One evacuee, Tasha Thomas-Naquin, is quoted in *The New Orleans Tribune* (Hurst 2011) as saying:

> Early after Hurricane Katrina, getting that taste of New Orleans required a little ingenuity and driving … When I was in Charlotte, N.C., Patton's sausage was one of the things I missed most and the DD smoke sausage [sic]—especially when I wanted to make gumbo, she says. I would take the ten hour ride all the way back home with my cooler in the back of my van and stock up on all my New Orleans foods I couldn't get in North Carolina—blue crabs, DD smoke sausage, Patton's hot sausage, alligator sausage and Louisiana seafood. If it wasn't for a lil' taste of home during those times I think I would have lost my mind. The food really brought me comfort.[5]

Thomas-Naquin's willingness to drive 10 hours or more with her cooler speaks to her need to stabilize herself with foods that are familiar and comforting but also that define her identity. Not only does food create a bond among community members despite their being apart, but also it links people across time and space. More than simply personal choice or objects of emotion, foods that provide a sense of security are "deeply embedded within larger social and cultural systems" (Locher et al. 2005: 277). So, it is significant that the comfort foods of New Orleans described by Thomas-Naquin herself are not the same as those shared by Blacks living in the American South

Food helps to maintain group cohesion, tying individuals to society. Though indirectly stated, Thomas Naquin makes the same point of differentiation between creole and Southern soul as Wayne Baquet, owner of Dizzy's Café in New Orleans:

> Oh there's definitely a difference between Creole food and soul food, without a doubt. Creole food is soul food. It's the soul food of New Orleans. But—there's a total difference between Creole cooking and soul food. Down home soul food cooking we don't do. We don't do chitterlings, okay. We don't do ox-tails. We don't do pig-tails, and we won't do those things. Those things are soul, down home Soul Food, you know. We—all the things we do are New Orleans Creole things. Like stuffed peppers, jambalaya, crawfish bisque, crawfish pies. So there's a distinct difference between what we do and what is strictly soul food. (Roahen 2007)

---

5   Thomas-Naquin is referring to Double D Smoked Sausage or D & D, a popular sausage for residents. Many businesses now specialize in shipping foods from the region directly to consumers.

Failure to recognize these similar yet different practices of ethnicity can lead to interethnic tensions. For example, going to New Orleans and expecting to find the same foods you might find in other parts of the south can lead to misunderstanding. This is not to say that you cannot find Southern soul food in the Bayou State. Quite the contrary. By virtue of its geographical location, Louisianan cuisines, both Creole—which borrows from the Native Americans, Mexican Aztecs, Europeans of French, German, Spanish ancestry, Africans, and West Indians—and Cajun—which derives largely from Acadia—obviously have come into contact with Southern foods. The point here is that food is related to social boundaries and is often accommodated by certain particularities.

In related fashion, the same foods can resonate differently depending upon the context and may be used—consciously and unconsciously—to communicate about themselves and their in-group status (Kalcik 1984: 54). The aromatic combination of onion, green bell pepper, and celery, combined with bay leaves, garlic (powder), paprika, and cayenne forms the basis for many Creole dishes from jambalayas—a common Louisianan dish eaten throughout the state that is similar to Spanish paella or West African Jolof rice in the sense that it is comprised of rice, meat, vegetables, seafood—to gumbos. Additionally, there is always an undeniable presence of a sauce of some sort—red or brown—that forms the roux. Knowing this may not mark you as an insider but may suggest that time has been taken to gain an awareness of the group's cultural practices. Rather than view these examples of ethnic differentiation as a deviations from the norm, it is more instructive to read them as varied performances of African American culinary identity.

Nyam means Eat ... Gullah Food Culture

> 'E teif me pinders n e hand een ain onrabel e mout!'
> 'What? I don't understand a word you are saying! Speak English!'
> 'She can't speak no English, teacher. She is one of those bad-talking, rice eating Geechees from South Carolina.'
> Laughter.
> The laughter hurt my ten-year old heart ... I longed to be back in the South Carolina Low Country where eating rice wasn't funny and the teacher would understand when I said a boy stole my peanuts and ran off without saying a word. (Vertamae Grosvenor, *Vertamae Cooks*)

Foodways encode social events and interactions so much that specific foods and customs often come to be associated closely with the groups that practice them. Similar to Louisianan cooking where gumbo is a signifier, ever-present in Gullah food culture is rice (Joyner 1984, Hess 1992, Carney 2001), a staple in preparation and consumption owing to its massive cultivation in the Lowcountry region. The geographic and cultural region located along South Carolina's coast, the

Lowcountry is part of the Sea Islands, a band of islands from Florida to Georgia and South Carolina on the coast of the Atlantic Ocean. During the eighteenth century, African slaves were brought to the region—primarily from West and Central Africa as well as the Caribbean—for their knowledge of cotton, rice, and indigo planting. More than other African American communities, the Gullah, as people in this region are known, have preserved their linguistic and cultural heritage, including their food traditions. Even though Gullah people use some of the same ingredients as those in other parts of the South, including Louisiana, they are distinctive in practice because they reflect some African culinary practices. This is due largely to their physical location along the coast and having had less interaction with white planters. Despite variations in the demographic structures and economic practices among the residents most of the denizens are "direct descendants of enslaved Africans" who worked on the islands as early as the seventeenth century (Beoku-Betts 1995: 539).

Perhaps the phrase that best encapsulates Gullah cultural foodways is slow cooking because historically local denizens farmed, gardened, fished, and crabbed and then cleaned, cut, chopped, and simmered.[6] Even though massive tourism on the island of Hilton Head and surrounding resorts has brought an influx of fast food restaurants, some residents have persisted in obtaining their foods from primarily local sources. This custom is aided in part by the presence of rural lifeways—some residents still live in areas inaccessible by bridges. This kind of food self-sufficiency necessarily gives birth to the passing on of traditions, particularly the art of cooking rice. More than a staple dish, rice in Gullah culture represents a rich heritage reflective of resilience and African traditions. Not only are many older residents able to distinguish various types of the grain but also they are conversant in explaining how the rice needs to be cooked, ways of doling out rice and water without measuring cups or spoons, the kinds of pots to use, and the amount of time in which the rice needs to be cooked. As Gullah resident, Carla Bates, testifies: "Many people feel if rice isn't cooked, they haven't eaten" (Beoku-Betts 1995: 543).

What both ties and distinguishes Gullah cooking from/to southern foods and "soul foods" as well as Creole cooking are the kinds of foods that are cooked, the ways they are cooked, and the seasoning combinations used, some of which are directly connected to West African cookery. Again, I turn to the study conducted on the Gullah people:

> One of these practices involves the selection, the amounts, and the combination
> of seasonings for food. These elements differentiate Gullah cooking practices

---

6   Slow food cooking is an international movement reestablished by Carlo Petrini in the late 1980s as a viable alternative to the rapidly expanding fast food industry. The goal is to preserver traditional and regional cuisines using foods gleans from the local ecosystem. For more on the Slow Food movement in the United States see http://www.slowfoodusa.org/index.php/slow_food/.

from those of other cultures, according to many women I interviewed. Although the Gullah identify certain foods as their own, such as Hoppin' John (rice cooked with peas and smoked meat), red rice, rice served with a plate of shrimp and okra stew, and collard greens and cornbread, the interaction between European American, Native American, and African American food systems in the South has carried these popular southern dishes across ethnic lines. One way in which Gullah women try to control cultural boundaries in their way of cooking these foods, as distinct from other southern practices, is to assert that although similar foods are eaten by others in the South, their style of preparation and the type of seasonings they use are different. Just as West African cooking is characteristically well seasoned with salt, pepper, onions, garlic, and smoked meat and fish, Gullah food is flavored with a combination of seasonings such as onions, salt, and pepper, as well as fresh and smoked meats such as bacon, pigs' feet, salt pork, and (increasingly) smoked turkey wings (to reduce fat content). (Beoku-Betts 1995: 547)

Here, Beoku-Betts illustrates a critical point about foods. They are flexible enough to perform the same or different functions. Rice, for example, in one locale is a dietary supplement, while in another it is a staple. Similarly, as Beoku-Betts (1995: 547) notes above, "although similar foods are eaten by others in the South, [Gullah] style of preparation and the type of seasonings they use are different."

This deliberate differentiation is, inevitably, part of the process of controlling Gullah's cultural boundaries and it is what separates this group of African American culinary practices from other groups. These variances, though seemingly subtle are actually stark because the same ingredients are made to be quite different in taste. The meat might be stewed longer, the okra less crunchy, the tomato less prevalent. An insider will expect the modifications. All of this reveals how foods are part of the process of cultural continuity as well as difference. African Americans may eat the same foods but we cook and eat them differently. This also reflects a culinary heterogeneity that is often unrecognized by outsiders.

Black Food Is/Black Food Ain't—Conclusion

One of the fastest growing instances of intermarriage in the United States is occurring between African men and African American women (Durodoye and Coker 2008, William-Forson 2010, McCabe 2011). As a result, more and more African American and West African meals are being combined. This new wave of Black heritage food is more visibly emphasized by patterns of transnational migration. While these migrations have always existed and are clearly reflected in the food cultures discussed in this chapter, the advent of ethnic markets throughout the United States enables more culinary incorporation between and among foods of African, Caribbean, Southeast Asian, and Latin American cultures. This realization undergirds the project of my student (Henry 2009: 9) that opens this chapter. Her final project worked to reconcile these admixtures from a food and race point of view:

> Being that my family is from both the United States and Guyana, during the holidays and other celebrations we usually have a wide array of foods at our table. My sister calls it a 'hybrid feast of all the best African American foods' ... In this way, food preserves family heritage and makes a link between the past and present. In the Caribbean American home food acts in the same way [as the African American home]; it ties individuals to and preserves their homeland and heritage as they participate in their traditional ways of consumption. (Henry 2009: 9)

From the souse to salt fish cakes, metemgee (a popular, traditional Guyanese dish) to cook-up rice (a one-pot Guyanese national recipe), sweet potatoes to collard greens, smothered chicken to the New Orleans Monday tradition of cooking Camillia brand red beans, Black food cultures in the United States encompass various culinary legacies. African Americans tell stories with the foods that they have harvested, prepared, sold, and consumed. The recipes passed down from generation to generation—whether on the porches of those residing on Martha's Vineyard or in neo-soul restaurants like San Francisco's Farmer Brown's—evoke more than the smells, sights, sounds, and tastes of foods associated with a single racial identity. They mark time and tell the fertile histories of a dynamic culinary cultural heritage of many ethnicities within one race.

References

Ball, C. (1858) 1970. *Fifty Years in Chains, or, The Life of an American Slave.* New York: Dover Publications.

Baraka, A. (LeRoi Jones). 1966. Soul Food. In *Home: Social Essays.* New York: William Morrow and Co, 121–3.

Barth, F. 1969. Introduction. In F. Barth (ed.), *Ethnic Groups and Boundaries,* (pp. 9–38). Boston, MA: Little, Brown and Company,.

Beoku-Betts, J. 1995. We Got Our Way of Cooking Things: Women, Food, and Preservation of Cultural Identity Among the Gullah. *Gender and Society,* 9(5), 535–55.

Brown, L.K. and K. Mussell, Introduction. In L. Keller Brown and K. Mussell (eds), *Ethnic and Regional Foodways in the United States: The Performance of Group Identity.* Knoxville: University of Tennessee Press, pp. 3–18.

Carney, J. 2001. *Black Rice: The African Origins of Rice Cultivation in the Americas.* Cambridge, MA: Harvard University Press.

Chai, S. 2005. Predicting Ethnic Boundaries. *European Sociological Review* 21(4), 375–91.

Chavis, S. 2006 Their Gullet: From Someone Else's Kitchen. (online) Available at: http://forums.egullet.org/index.php?/topic/87448-their-food/ (Accessed 28 February 2008).

Deetz, J. 1996. *In Small Things Forgotten.* New York: Anchor Books.

Di Giovine, M. 2010. La Vigilia Italo-Americana: Revitalizing the Italian-American Family Through the Christmas Eve "Feast of the Seven Fishes." *Food and Foodways* 18, 181–208.

Durodoye, B and Coker, A. 2008. Crossing Cultures in Marriage: Implications for Counseling African American/African Couples. *International Journal for the Advancement of Counseling* 30(1), 25–37.

Ferguson, L. 1992. *Uncommon Ground: Archaeology and Early African America, 1650–1800.* Washington: Smithsonian Institution Press.

Franklin, M. 2001. The Archaeological and Symbolic Dimensions of Soul Food: Race, Culture and Afro-Virginian Identity. In C. Orser (ed.), *Race and the Archaeology of Identity.* Salt Lake City: University of Utah Press, 88–107.

Gabaccia, D. 1998. *We Are What We Eat: Ethnic Food and the Making of Americans.* Cambridge, MA: Harvard University Press.

Giddens, A. 1991. *Modernity and Self-identity. Self and Society in the Late Modern Age.* Cambridge: Polity Press.

Gordon, R. and Neville, M. (Directors) 2007. *Respect Yourself: The Stax Records Story* [DVD]. United States. Tremolo Productions

Grosvenor, V. 1996. *Vertamae Cooks in America's Family Kitchen.* San Francisco: Kqed Books.

Hall, R. 2007. Food Crops, Medicinal Plants, and the Atlantic Slave Trade. In A. Bower (ed.), *African American Foodways: Explorations of History and Culture.* Urbana: University of Illinois Press, pp. 17–44.

Hall, S. 1997. Old and New Identities, Old and New Ethnicities. In A. King (ed.), *Culture, Globalization and the World-System: Contemporary Conditions for the Representation of Identity.* Minneapolis: University of Minnesota Press, pp. 41–68.

____. 1990. Cultural Identity and Diaspora. In J. Rutherford (ed.), *Identity: Community, Culture, Difference.* London. Lawrence & Wishart, pp. 222–37.

Harris, J. 1995. *The Welcome Table: African American Heritage Cooking.* New York: Simon & Schuster.

Henry, M. 2009. "Guyanese/African American Foods." Paper written for course—Black Class: From the Harlem Renaissance to Hip Hop (University of Maryland College Park).

Hess, K. 1992. *The Carolina Rice Kitchen: The Africa Connection.* Columbia: University of South Carolina Press.

Heuman, G and Walvin, J. (eds) 2003. *The Slavery Reader.* London and New York: Routledge.

Hurst, W.B. A Lil' Taste of Home: On the Sixth Anniversary of Hurricane Katrina New Orleans Cuisine Reigns Supreme. The New Orleans Tribune (online 2011) Available at: http://theneworleanstribune.com/wilmarine.htm (Accessed 1 March 2012).

Joyner, C. 1984. *Down by the Riverside: A South Carolina Slave Community.* Urbana: University of Illinois Press.

Kalcik, S. 1984. Ethnic Foodways in America: Symbol and the Performance of Identity. In L.K. Brown and K. Mussell (eds), *Ethnic and Regional Foodways in the United States: The Performance of Group Identity.* Knoxville: University of Tennessee Press, pp. 37–65.

Lamont, M. and Molnár, V. 2001. How Blacks Use Consumption to Shape Their Collective Identity: Evidence from African-America Marketing Specialists. *Journal of Consumer Culture* 1, 31–45.

Latshaw, B. 2009. Food for Thought: Race, Region, Identity, and Foodways in the American South. *Southern Cultures*, 15(4), pp. 106–28

Levenstein, H. 1988. *Revolution at the Table: The Transformation of the American Diet.* New York: Oxford University Press.

Locher, J., Yoels, W., Maurer, D. and van Ells, J. 1999. Comfort Foods: An Exploratory Journey Into The Social and Emotional Significance of Food. *Food and Foodways*, 13(4), 273–95.

McCabe, K. U S in Focus: African Immigrants in the United States, Migration Policy Institute (online July 2011) Available at: http://www.migrationinformation.org/USfocus/display.cfm?ID=847#5 (Accessed March 3, 2012).

McKee, L. 1999. Food Supply and Plantation Social Order: An Archaeological Perspective. In T. Singleton (ed.), *"I, Too, Am America": Archaeological Studies of African-American Life.* Charlottesville: University Press of Virginia, pp. 218–39.

Morgan, P.D. 1998. *Slave Counterpoint: Black Culture in the Eighteenth-Century Chesapeake and Lowcountry.* Chapel Hill: University of North Carolina Press for the Omohundro Institute of Early American History and Culture, Williamsburg, Virginia.

Nagel, J. 1994. Constructing Ethnicity: Creating and Recreating Ethnic Identity and Culture. *Social Problems*, 41: 152–76.

Nettles, K.D. 2007. Saving Soul Food. *Gastronomica*, 7(3), 106–13.

Okamoto, D. 2003. Toward a Theory of Panethnicity: Explaining Asian American Collective Action. *American Sociological Review,* 68(6), pp. 811–42

Omi, M. and Winant, H. 1994. *Racial Formation in the United States: From 1960s to 1990s.* New York: Routledge.

Opie, F.D. 2008. *Hog and Hominy: Soul Food from Africa to America.* New York; Columbia University Press.

Paige, H. 1999. *Aspects of African American Foodways.* Southfield, MI: Aspects Publishing Company.

Pulsipher, L. 1990. They Have Saturdays and Sundays to Feed Themselves: Slave Gardens in the Caribbean. *Expedition* 32(1), 24–33.

Roahen, S. 2007. Wayne Baquet, Sr. (online: The Southern Gumbo Trail, An Oral History Project). Available at: http://www.southerngumbotrail.com/baquet.shtml (Accessed 1 March 2008).

Samford, P. 1970. The Archaeology of African-American Slavery and Material Culture. *The William and Mary Quarterly*, 53(1), 87–114.

Singleton, T, (ed.). 1985. *The Archaeology of Slavery and Plantation Life*. Florida: Academic Press.

Tipton-Martin, T. 2008. Who is Mandy? The Austin Chronicle (online). Available at http://www.austinchronicle.com/food/2006-02-24/341982/ (Accessed 28 February 2008).

Walsh, L. 1997. *From Calabar to Carter's Grove: The History of a Virginia Slave Community*. Charlottesville: University Press of Virginia, 1997.

Waters, M. 1994. Ethnic and Racial Identities of Second-Generation Black Immigrants in New York City. *International Migration Review* 28, 795–820.

Whitehead, T. 1992. In Search of Soul Food and Meaning: Culture, Food, and Health. In H. Baer and Y. Jones (eds), *African Americans in the South: Issues of Race, Class, and Gender*. Athens: University of Georgia Press, pp. 94–110.

Wilk, R. 1999. 'Real Belizean Food': Building Local Identity in the Transnational Caribbean. *American Anthropologist*, 101(2), 244–55.

Williams-Forson, P. 2010. Other Women Cooked for My Husband: Negotiating Gender, Food, and Identities in an African American/Ghanaian Household. *Feminist Studies* 36(2), 435–61.

____. 2006. *Building Houses Out of Chicken Legs*. Chapel Hill: University of North Carolina Press.

Witt, D. 1998. Soul Food: Where the Chitterling Hits the (Primal) Pan. In R. Scapp and B. Seitz (eds), *Eating Culture*. Albany: State University of New York Press, pp. 258–87.

____.1999. *Black Hunger: Soul Food and America*. NY: Oxford University Press.

Yentsch, A. 1992. Gudgeons, Mullet, and Proud Pigs: Historicity, Black Fishing, and Southern Myth. In M. Beaudry and A. Yentsch (eds), *The Art and Mystery of Historical Archaeology*. Boca Raton: CRC Press, pp. 283–314.

____.1995. Hot, Nourishing, and Culturally Potent: The Transfer of West African Cooking Traditions to the Chesapeake. *Sage* 9(2), 15–29.

Chapter 7

Caldo De Piedra and Claiming Pre-Hispanic Cuisine as Cultural Heritage

Ronda L. Brulotte and Alvin Starkman

In November 2010, Mexican cuisine, represented by the traditional foodways of the west-central state of Michoacán, was inscribed on UNESCO's list of the Intangible Cultural Heritage of Humanity (Mazatán-Páramo 2010). However, decades before then, the southern Mexico state of Oaxaca had been internationally recognized for its cuisine, and in the view of many, locals and outsiders alike, is the state with the best food in the country, largely because of Oaxaca's rich and diverse culinary heritage.

Oaxaca de Juárez, the state capital located in the middle of a series of central valleys, has at least six cooking schools, with additional classes offered by many of the city's Spanish language schools. Oaxaca's central valleys and surrounding districts have become synonymous with uniquely Oaxacan foods such as *chapulines* (fried and seasoned grasshoppers), *tlayudas* (oversized tortillas spread with a layer of beans and pork fat, vegetables, cheese, and meat), *quesillo* (Oaxacan string cheese), *gusanos* (larvae that infest the agave plant used to make mezcal, and are utilized as a salsa ingredient or when ground with salt and chile as a chaser), and the most renowned of all, its moles: sauces made with up to 35 ingredients including dried and fresh chiles, fruits, seeds, nuts, spices and even chocolate.

Oaxaca has also become a major center for visiting national and international chefs, food critics and journalists, "foodies," tourists, and even anthropologists (see Haines and Sammells 2010) who have heard about Oaxaca's culinary heritage and want to experience it firsthand. Oaxacan cuisine is regularly featured in newspapers, magazines, and online blogs throughout the world, and is the topic of several English-language cookbooks (Kennedy 2010, Martínez 1997, Trilling 1999). And because tourism is one of the state's top three revenue producers (along with petroleum and migrant remittances), significant efforts are made at both state and federal levels to disseminate information about gastronomy in Oaxaca. In February 2012, in an effort to build on the UNESCO designation, SECTUR (the federal tourism secretariat) designated 18 gastronomic routes throughout Mexico, naming a route centering upon Oaxaca, *Los mil sabores del mole* (The thousand flavors of mole). At the state level, SEDETUR (the Secretariat of Economic Development and Tourism) is working to promote local food festivals in communities throughout Oaxaca. Tourist visitors can attend festivals honoring *tejate* (a pre-Hispanic drink made from a base of maize and cacao) in

San Andrés Huayapam, mezcal (a distilled agave spirit) in Santiago Matatlán, and wild mushrooms in Cuahimoloyas, to name a few. Meanwhile, the town of Santa María El Tule, a short drive east of Oaxaca city, claims to be the "the birthplace of the empanada"—a title proudly announced at the town entrance. More recently at the end of 2013, the federal and state governments, in conjunction with a private, Oaxaca-based foundation, allocated 30 million pesos (over two million U.S. dollars) to create a Gastronomic Center in the capital city as part of the ongoing development of Oaxaca's tourist sector.

The competition to capture a share of Oaxaca's growing culinary market has intensified within the private market as well, and local restaurateurs increasingly seek to define their offerings against others'. Today there is even a handful of eateries in the capital advertising *cocina de autor*, an international culinary movement that stresses individual chefs' innovative and artistic cuisine inspired by regional ingredients and dishes.

Still, for most Oaxacan chefs and restaurant owners, there are no carefully guarded secrets, nor claims to proprietary culinary rights. But César Gachupín Velasco, an Oaxacan restaurateur, and his town elders view things differently. Gachupín was born in San Felipe Usila, a municipality of approximately 11,500[1] residents, many indigenous Chinantec-speakers, in the district of Tuxtepec, a five- to six-hour drive from the capital. In 1996 he opened the first of three restaurants named Caldo de Piedra on main street of his village, but later moved his business to the outskirts of the city of Oaxaca de Juárez, where he went on to open restaurants in 1999 and 2006. As the Spanish name implies, the restaurants' signature dish is cooked by heating rocks (*piedras*) and placing them in a half gourd with seasoned broth (*caldo*) and fish or shellfish. According to Gachupín, this cooking method together with the dish's components are the cultural property of his community, and anyone who cooks such a meal without the permission of his village assembly is stealing a protected recipe without color of right, no different than if he or she were making and selling pirated CDs or DVDs.

Using the case study of Caldo de Piedra, this chapter examines the tensions and contradictions that arise in claiming a culinary practice as the cultural property of a specific indigenous community while at the same marketing it under the umbrella of Mexico's UNESCO Intangible Cultural Heritage designation to a global clientele. We demonstrate that although regional, national, and even "world" heritage claims are political and operate from the top-down, heritage claims around food are ultimately understood, valued, and really "worked out" by social actors in their everyday practices and interactions on the ground.

---

1 Population statistic reported by the Istituto Nacional de Estadística y Geografía (INEGI), Censo de Población y Vivienda, 2010 (http://www.inegi.org.mx/sistemas/mexicocifras/default.aspx?e=20&mun=136&src=487, accessed 23 June 2012).

Oaxaca's Indigenous Culinary Heritage and Caldo de Piedra

Oaxaca is historically not only one of Mexico's economically poorest states; it is also one of the most ethnically diverse, with at least 15 different indigenous languages spoken. The region remained by and large isolated from the cultural sphere of influence of the United States, and even from the country's capital, Mexico City, at least until the 1940s when the Pan American Highway reached Oaxaca, connecting it to other regions. Oaxaca's inaccessibility resulted from its ruggedness, and consequent geographical segregation; its climatic, physical, and biological diversity coupled with its multiplicity of cultural traditions stood Oaxaca apart from other states in Mexico.

Oaxaca's culinary heritage, however, did not grow in isolation. The Spanish arrived in what is now the present day state in 1521. Subsequently other European nations, France in particular, played an important role in contributing to the history and development of Oaxacan society, including its cuisine. Oaxacan gastronomy is thus a melding of native ingredients and indigenous cooking techniques and instruments with processes and foodstuffs introduced into the state by European invaders and settlers (see Pilcher 1998). In fact, most visitors who make a pilgrimage to Oaxaca to learn about its moles do not realize that several key ingredients in its famed *mole negro* (black mole) are not native to Mexico at all, but rather the result of Europeans first importing, then cultivating particular foodstuffs; almonds, raisins, sesame seeds, and the meats typically used for the dish (pork or chicken) all came from elsewhere. Other commonly used "local" ingredients such as peanuts, cinnamon, anise, clove, cumin and peppercorn likewise are not indigenous to Oaxaca, or Mexico for that matter.

Despite these historical details, Oaxacans of all socioeconomic strata increasingly take pride in the fact that local food maintains direct links to indigenous ingredients and preparations. Even Oaxacan elites, who a generation ago would have eschewed corn tortillas in favor of wheat bread and chosen imported scotch and brandy over locally distilled mezcal to serve at lavish parties, are "discovering" Oaxaca's culinary heritage, now so actively promoted within the state. The region has also recently come to the attention of foodies and proponents of sustainable food systems alike for the local consumption of insects (notably the abovementioned grasshoppers and *gusanos* as well as *chicatanas,* a variety of flying ant), which were an important source of protein in a pre-Hispanic Mesoamerican diet that included few animals. Recent archaeological work at Guilá Naquitz cave near Mitla in Oaxaca's Central Valleys has yielded one of the earliest macro specimens of domesticated maize in Mexico, dating to around 4300 B.C. (Kennedy 2010, Benz 2005). Corn, along with chile, beans, and squash—pillars of the pre-Hispanic diet—still figure prominently in the majority of Oaxacans' daily caloric intake, particularly in rural areas. Likewise, in Diana

Kennedy's[2] recent culinary tome, *Oaxaca al Gusto* (2010), the author and culinary expert stresses seemingly infinite diversity of these native ingredients produced within the various microclimates found throughout the state.

Folklorist Lucy Long has described food as both "a destination and vehicle for tourism" (Long 2004: 2). In the first instance, food itself is the primary motivator for travel to particular towns or regions. Wine lovers from around the globe traveling in Europe make a pilgrimage to several growing regions in France, while cheese aficionados select either Normandy for Camembert or the Île-de-France for brie. Scotch drinkers vacationing in the UK invariably make a point of visiting the regions in Scotland from where their favorite whiskies emanate, often including Islay's rocky southern shore to pay homage to Lagavulin's pungent, peaty smoky flavor produced by the climate's unique malt. At the same time, food may function as a vehicle for diversified touristic ventures. People may in part go to places to experience the food, but these culinary interests synergistically interact with other opportunities provided by the local tourism complex; bicycle tours, visits to museums and other cultural attractions, and shopping typically round out the itineraries of culinary tourists, generating revenue in multiple venues.

Culinary tourism is already a burgeoning industry throughout the world, and the current attention lavished on indigenous Oaxacan foodways goes hand-in-hand with the development of a larger, well-developed heritage tourism industry that is heavily focused on folk arts and archaeology (Brulotte 2012). Culinary enthusiasts in Oaxaca actively search out villages with a reputation for a particular quality of mezcal or market stall or restaurant known for a particular local foodstuff, particularly if it has been featured on a popular food program, such as celebrity chef Anthony Bourdain's "A Cook's Tour." With its proprietors claiming to serve "the only pre-Hispanic dish declared by UNESCO as Intangible Cultural Heritage" (www.caldodepiedra.com), Caldo de Piedra restaurants are well placed to capitalize on the culinary travel trend, particularly among what Lisa Heldke (2003) terms "food adventurers," those seeking exotic, authentic food experiences outside their everyday world.

Gachupín's flagship Caldo de Piedra is located on the highway between Oaxaca de Juárez and Santa María el Tule, the self-proclaimed "birthplace of the empanada" and home to a popular attraction: a 2,000-year old cypress, a tree with allegedly the largest girth in the world. Tourists visiting the state capital, regardless of whether they are interested in botany and ecology, ruins and pre-history, craft

---

2   Diana Kennedy is considered to be one of the world's foremost experts on Mexican cuisine. She has written numerous books on Mexican food, and her 1972 book *The Cuisines of Mexico* was among English-language cookbooks dedicated to educating North American audiences about "authentic" regional foods. In 1981 Kennedy was awarded the Order Of the Aztec Eagle (one of the highest national awards) by the Mexican government for her contributions to the documentation of Mexican cuisine. British-born, Kennedy has lived much of her adult life in Mexico and currently resides in the state of Michoacán.

villages, colonial architecture, or cuisine, invariably visit Santa María el Tule. Accordingly, the vast majority of tourists pass right in front of Caldo de Piedra. And if they do not drive by Caldo #1, they come across the second branch, which opened in late 2011; it is located along the other popular tour route, leading to many of the best-known craft villages in Oaxaca's central valleys, including San Bartolo Coyotepec, a town famous for its black pottery.

Upon entering Caldo de Piedra, one walks into a large, thatched palm leaf roofed, semi-open-air restaurant with non-descript furnishings. The establishment has a limited menu focusing on caldo de piedra with fish, shrimp, or a combination of the two, as well as few other more commonplace dishes such as quesadillas and shrimp adobado (neither of which are explicitly promoted by the restaurant as forms of cultural heritage). Family members wait on customers, and if it is their first time, they are invited to the back of the restaurant to watch the caldo preparation, usually carried out by the owner or his son.

Visitors watch, some with digital cameras or camera phones in hand, as a staff member sets out on a counter in front of a hearth with flaming logs, a half, round-bottomed gourd, which rests on a *rodete* (an individual river reed ring) to ensure that the bowl does not tip (see Figure 7.1). Gachupín is usually close by, working on another counter filling blenders with a tomato-based broth, then herbs and spices. He switches them on, and then half fills each gourd with the seasoned liquid. Finally, he uncovers plastic containers with fish and shrimp, and places in each gourd a healthy amount of one or the other or some of each.

The public display of cooking then begins. Gachupín picks up a rustic looking, tree-branch tong, removes hot rocks—transported from his home region—from the hearth one at a time, placing two in each gourd. The broth immediately begins to bubble. A couple of minutes later, he removes the rock and places another in each bowl, rejuvenating the boiling (see Figure 7.2). Each gourd is then brought to the table and placed on a *rodete* in front of diners; the cooking continues for a few more minutes until the seafood is cooked through.

While Gachupín ceremoniously prepares caldo de piedra, he briefly expounds about his village, and then informs his audience about the origins of the tradition, and his exclusive right to replicate it as a community member of San Felipe Usila. His narration of the dish's history projects back into a pre-Conquest past and to set of practices that he imagines forms the basis of the more recent preparations with which he is familiar. At the center of his discourse is the human capacity to transform the most basic elements at hand—rocks, fire, and water—into a means of sustenance that is not only biological but also symbolic; the products of nature are transformed into "culture" even through the most rudimentary human action.

In a 2012 interview Gachupín offered a more detailed account of his family's farming background and indigenous heritage (a native speaker of *chinanteco*, he only mastered Spanish in his third year in school), as well as caldo de piedra's origins and how he came to translate the dish for an urban audience. He spoke of "rescuing history," insisting that caldo de piedra is an "ancient, totally millenary" cooking practice that was uniquely developed by fishermen in his community to

Figure 7.1   Restaurant visitors watching the preparation of caldo de piedra

Source: Photo by Ronda Brulotte.

Figure 7.2   Cooking caldo de piedra by adding heated rocks to the ingredients in gourd bowls

Source: Photo by Ronda Brulotte.

Figure 7.3   English-language poster in Oaxaca City advertising the Caldo de Piedra restaurant

Source: Photo by Ronda Brulotte.

prepare and consume fresh catches of fish at the Papaloapan River's edge during the spring season. According to him, this ritual is grounded in a historically gendered division of labor. Women bathe and wash clothes in the river, but it is only adult men who fish and prepare the caldo, with the finished soup finally being offered to women and children.

The communal nature of caldo de piedra's preparation and consumption has apparently fostered a sense of ownership over the dish within San Felipe Usila itself. In order to open his Oaxaca City restaurants, Gachupín maintained that he had to first request permission from a tribunal of community elders, who ultimately conferred their blessing. He thus interpreted his entrepreneurial endeavor as having motivations beyond economic gain, even as the restaurants' polished website and other marketing devices such as glossy posters (see Figure 7.3) and newspaper ads suggest otherwise. "You're not just doing business, you're transmitting a culture," he stressed.

By his own admission, Gachupín had not been overly successful in courting an urban Oaxacan clientele, whom he believed to be less curious about food traditions outside of their local purview. He instead aimed to attract tourists, and reported that he had received national visitors from Campeche, Mexico City, Puebla and a number of other Mexican states. Yet he found foreigners to be the most appreciative and inquisitive of his clientele. Some came simply to eat and see the process, while others were interested in documenting caldo preparation in some form or another. The restaurant reported hosting crews from the Discovery Channel and National Geographic, as well as students and faculty from U.S. and Canadian universities.

Beyond Authenticity

It is tempting to try and discern whether the culinary performance one encounters at the Caldo de Piedra restaurants is yet another example of what historians Eric Hobsbawm and Terence Ranger (2003) famously described as an "invention of tradition," a practice that is considered traditional but in reality is a much more recent creation, often constructed to serve a specific ideological (or economic) end. The well-rehearsed speech asserting caldo de piedra's ancient origins and stressing the owner's wish to safeguard his community's food heritage, along with the stylized presentation of the dish itself—far from the river's edge with a captive, paying audience—all point to a self-conscious crafting of tradition.

The restaurants' use of certain ingredients and preparation techniques are inconsistent with the story of the original dish. Indigenous women in traditional garb grind corn on a *metate* (grinding stone) to make tortillas, but an electric blender churning tomato and herbs for the broth drowns out the deliberate sound of stone against stone. In the district of Tuxtepec and other regions throughout Oaxaca, fish are caught and freshwater seafood (*langostinos,* a cross between large shrimp and crawdads) is harvested at minimum on a seasonal basis from inland lakes and dams, rivers and streams. One might ask, if strict adherence to community tradition is integral to the dish, then why, 12 months a year, does Caldo de Piedra serve only shrimp and red snapper, seafood caught from the Pacific Ocean?

Aside from caldo de piedra, the recent addition to the menu of shrimp adobado is also inconsistent with the restaurant's marketing plan. Adobo is a seasoning mix, used in the Iberian Peninsula before the arrival of the conquistadors in the New World. Historically it has been used to flavor dishes throughout Latin America and the Caribbean, as well as in the Philippines (in the latter, it is most often considered a complete dish, referred to and acknowledged as a, if not *the,* national plate). Furthermore, apart from using saltwater shrimp rather than anything from freshwater and creating a recipe using flavorings of herbs and spices native to San Felipe Usila, or even Mexico, Gachupín elected to use adobo, Mexican incarnations of which include black peppercorns and/or cumin, neither of which is indigenous to the country. Yet even some of Oaxaca's "traditional" primary meat proteins prepared using acknowledged widespread ancient cooking techniques are not native to Oaxaca. For example, the ceremonial slow cooking of *borrego* (sheep) in an in-ground pit oven, features an animal that was not even introduced into the central valleys of Oaxaca until 1535, by a Dominican. Yet no one has suggested that *barbacoa de borrego* is deserving of cultural protection.

Most anthropologists today are much less interested in finding some true or unadulterated cultural content—something that needs "saving"—than in understanding what Pierre Bourdieu (1993) termed the contemporary "field of cultural production," in which ideas about culture are produced, circulated, and re-worked within networks of individuals, groups, and institutions. Bourdieu saw this field not only as a hierarchical, interdependent system of social actors but also as an arena of ongoing "struggles tending to transform or conserve this field of forces" (Bourdieu 1993: 30). Anthropologist Michael Di Giovine (2009) elaborates on this model, designating the "field of heritage production" as the arena where the active positioning and position-taking by stakeholders takes place; conflict and tension is invariably part of the process of the working out of the sites and practices that come to be collectively understood and accepted as "heritage." Within this framework, debating the authenticity of the products offered at Caldo de Piedra restaurants or their continuity from pre-Hispanic times misses the point. As Barbara Kirshenblatt-Gimblett reminds us, people "are not only cultural carriers and transmitters ... but also agents in the heritage enterprise itself" (Kirshenblatt-Gimblett 2004: 58). It is more fruitful, then, to examine the manner in which caldo de piedra is being promoted as the culinary heritage of a particular indigenous community as well as the repercussions of claiming it as cultural property in need of safeguarding.

In the same 2012 interview, when confronted with the suggestion that indigenous peoples in diverse parts of the world have been preparing meals at rivers' edges for generations by heating rocks and placing them in whatever kind of receptacle was available, Caldo de Piedra's owner was steadfast and to the point: "Someone from there must have come to San Felipe Usila or a neighboring community and copied the idea."

As with other Oaxacan cultural goods and practices that hold economic potential, for tourism or otherwise, appropriation is often the flip side of promotion in the name of cultural safeguarding. On its website and other marketing materials, Caldo de Piedra restaurants claim to serve the only pre-Hispanic dish included on UNESCO's list of Intangible Cultural Heritage of Humanity. Branding one's product as UNESCO heritage obviously has its benefits; tourists flock to Oaxaca's other UNESCO-declared sites: the city's historic colonial downtown and the nearby Monte Albán archaeological zone. It follows that they would seek out other UNESCO approved cultural venues. The only problem is that Caldo de Piedra is not actually on the UNESCO list. What *is* recognized as UNESCO Intangible Cultural Heritage is "traditional Mexican cuisine-ancestral, ongoing community culture, the Michoacán paradigm" (UNESCO 2010). While acknowledging the food practices of the west-central state of Michoacán, the UNESCO designation more generally recognizes a multi-faceted Mesoamerican alimentary system that is

> founded on corn, beans and chili; unique farming methods such as milpas
> (rotating swidden fields of corn and other crops) and chinampas (man-made
> farming islets in lake areas); cooking processes such as nixtamalization (lime-

hulling maize, which increases its nutritional value); and singular utensils including grinding stones and stone mortars. Native ingredients such as varieties of tomatoes, squashes, avocados, cocoa and vanilla augment the basic staples. (UNESCO 2010)

At the time of this writing, a group of chefs and food advocates from Oaxaca were seeking to secure a similar designation for their state, but to date, no one dish from Mexico has been singled out as UNESCO Intangible Cultural Heritage.

Why then does Gachupín claim UNESCO protection when technically there is none and exclusivity over a cooking technique that likely has developed in different parts of the world over millennia, using non-local ingredients and modern equipment? Barbara Kirshenblatt-Gimblett (1995) has argued that heritage is a "value-added" industry that works in collaboration with tourism. Labeling as heritage goods or practices those which have ceased to be valuable or were never economically productive confers value on them, and "it does this by adding the value of pastness, exhibition, difference, and where possible indigeneity" (Kirshenblatt-Gimblett 1995: 370). Caldo de Piedra notably traffics in all these forms of value creation, with its origin narrative that locates it in a remote but unspecific pre-Hispanic past, the performance of its preparation within the restaurant setting, and established links to a particular Oaxacan indigenous community. The assertion of the UNESCO Intangible Cultural Heritage "brand" may be interpreted as a synergistic outgrowth of this process.

The audience that Caldo de Piedra hopes to draw is obviously not located within the community of San Felipe Usila. Rather, it courts a public seeking out the culinary unfamiliar presented in a safe, accessible venue: "hygienic"—not in a market stall but in a well-maintained restaurant space—and within the city limits rather than a difficult six-hour drive into the mountains. Caldo de Piedra's ideal consumer represents an important, growing segment of Oaxaca's tourist demographic: international travelers who hail from the U.S., Canada, and Western European countries; are generally well educated; and occupy upper middle- to high-income brackets. These travelers are attracted to Oaxaca precisely because they perceive it to be a repository of "genuine" indigenous culture, past and present—travelers who, not coincidentally, are also those who are most likely to place value upon UNESCO branding of heritage around the world and to have the resources to visit UNESCO designated sites. Caldo de Piedra, then, is not a culinary anomaly within Oaxaca's cultural tourism market. On the contrary, it may be more appropriately read an indicator of what is to come, as Oaxaca continues its transformation into an upscale gastronomic destination and enterprising locals attempt to gain a market share. Food stalls and street carts will undoubtedly persist, but increasingly alongside organic farmers markets, celebrity chefs, self-described "artisanal" food producers, and English-language cooking schools and market tours.

Competing Claims

Casa Crespo is a high-end downtown Oaxaca restaurant that is representative of the culinary gentrification described above, perhaps in an even more overt way than its competitor Caldo de Piedra. Located in a colonial building across from Santo Domingo Church, one of Oaxaca's most iconic landmarks, Casa Crespo is owned by local proprietor Oscar Carrizosa. The bilingual Carrizosa initially started out running a bed and breakfast, a business that eventually evolved into a restaurant, cooking school, and outlet for artisanal chocolates and mezcals. Casa Crespo has received endorsements from *Bon Appétit* magazine, the *New York Times*, and the *Washington Post*, all of which are mentioned on its website (www.casacrespo.com). The prices at Casa Crespo are equivalent to what one might find at urban eatery in the United States, with plates averaging around $15–18 USD. But in Oaxaca, where the legal minimum wage is 67 pesos (just over $5 USD) per day and the majority of the population is economically disadvantaged, the restaurant and cooking school are prohibitively expensive for all but the city's elite and out of town visitors.

Casa Crespo's food offerings reflect the gastronomic traditions of all of the state's regions, and Carrizosa began serving caldo de piedra (using smaller peeled shrimp and preparing it tableside) several years ago, a decade after Gachupín opened the doors to his first Caldo de Piedra restaurant. On the menu, the dish is identified simply as a "Pre-Hispanic soup of shrimp and fish cooked with hot river stones," with no mention of it as belonging to a specific region, ethnic group, or village.

Casa Crespo's duplication of the dish sparked something of a culinary war between the two restaurateurs. In fact according to Gachupín, the San Felipe Usila assembly instructed him to retain a lawyer to write a cease and desist letter to Casa Crespo. Carrizosa reported that a lawyer in fact showed up at the restaurant to discuss the matter, then walked away after a friendly discussion between the two concluded. But Caldo de Piedra's website suggests that Gachupín has not let the matter drop entirely, offering a pointed response to those who would try to appropriate the dish to their own ends. It specifically warns visitors to not be taken in by false imitations of the original dish:

> After some time, various people outside of our culture have tried to imitate the process of making caldo de piedra, ignoring the strong cultural and prehispanic meaning that accompanies its preparation.
> Take care with false imitations of caldo de piedra. We are the people of Usila, the town where this dish originated and we only have two locations. (http://caldodepiedra.com/cuidado.html, accessed August 12, 2012, authors' translation)

Carrizosa, on the other hand, stood by his decision to serve caldo de piedra at his restaurant even though he has no connection to San Felipe Usila. When interviewed, he noted that he had come across anthropological literature citing cooking methods similar to caldo de piedra documented in the Canadian north and elsewhere. He also recalled a 1970s Latin American travel magazine that highlighted a similar caldo de piedra allegedly prepared by indigenous Mayan communities in Guatemala (closer to home, the authors have heard Oaxacans who were raised in different rural districts recount being out in the fields at lunchtime, going to the river, fishing, and preparing their meals in a manner akin to Gachupín's caldo de piedra).

While seemingly insignificant, the competing claims to the preparation of caldo de piedra points to an acceleration of the heritagization of culinary forms and practices in conjunction with the promotion of Mexico's massive tourism industry—ranked among the world's top 10 largest by the U.N. World Tourism Organization.[3] Food was somewhat of a latecomer to UNESCO's list of cultural heritage to be recognized or in need of safeguarding, but it is notable that Mexico was among the first nations in the world, in 2010, to have its gastronomic traditions recognized by the UNESCO international committee (the "gastronomic meal of the French" and the "Mediterranean diet," the latter a joint bid by Spain, Greece, Italy, and Morocco, also appeared on the Intangible Cultural Heritage list that same year).

At first glance, it is not entirely clear why the state of Michoacán was selected as the model for a complex set of Mesoamerican food practices that may be found throughout other regions of Mexico. However, upon closer examination of the original nomination file for Mexico one finds not only a joint effort by NGOs and several federal agencies, including the General Office of Popular Cultures (CONACULTA) and the National Institute of Anthropology and History (INAH), but also the state of Michoacán's tourism secretariat (SECTUR).[4] The slideshow and other photos showcasing the food traditions of Michoacán on UNESCO's website are provided by SECTUR as well. In other words, Mexico's bid to have its traditional cuisine included on the UNESCO Intangible Cultural Heritage list was necessarily conceived with its potential for tourism development in mind.

As mentioned above, caldo de piedra is not officially recognized by UNESCO yet that does not deter its proprietors from partaking in the status and potential economic advantage afforded by a UNESCO Intangible Cultural Heritage designation. But as anthropologist Lisa Breglia (2006) effectively illustrates in her study of archaeological heritage sites in Mexico's Yucatán Peninsula, the heritagization process, of which UNESCO branding is the example par excellence, produces a profound ambivalence among the various stakeholders involved. Intrinsic to the idea of heritage—with UNESCO purporting to promote and protect "World Cultural Patrimony"—is that it at once belongs to everyone, while belonging to no one in particular. Conflicts often arise when competing groups

---

3   Further details regarding these rankings may be found on the UNWTO's website at http://www.unwto.org.

4   The original application may be found at http://www.unesco.org/culture/ich/index. php?lg=en&pg=00011&RL=00400, accessed August 10, 2012.

claim special rights or exclusive access to heritage sites that attempt to override the assumed collective ownership implied within the notion of world heritage itself. However, unlike the archaeological monuments discussed by Breglia (the iconic Chichen Itza site and lesser-known Chunchucmil), culinary practices such as caldo de piedra are less easily fixed in time and space. Given the nature of its preparation, it may in fact be impossible to know exactly when and where the dish first arose, or if it simultaneously developed in multiple locations throughout pre-Hispanic Mesoamerica (or the rest of the world for that matter). To date, there is no archaeological evidence to suggest that the cooking technique first arose in the location today occupied by the municipality of San Felipe Usila.

There already exists in Mexico a legal mechanism, known in Spanish as *marca colectiva* (collective brand), to protect certain rural industries and products. Jointly coordinated by the Secretariat of Agriculture, Livestock, Rural Development, Fisheries, and Food (SAGARPA) and the Mexican Institute of Industrial Property (IMPI) in conjunction with state governments, officially registered *marcas colectivas* are meant to authenticate locally made products and prevent the proliferation of copies or imitations that undermine their economic viability. In 2005, Mexico awarded the first *marca colectiva* for an artisanal food product: traditionally produced *cotija* cheese from the Jalmich region along the Jalisco/Michoacán border. The model is similar to the Protected Designation of Origin status in Europe, which stipulates that products like Champagne and Roquefort cheese may be produced only in specific geographic regions, and in fact, may be accompanied by Mexican federal denomination of origin status. In 2010 in Oaxaca, women from 23 communities successfully secured a denomination of origin and *marca colectiva* under the name *"Mi Querencia"* for *tlayudas*, large, baked corn tortillas. *Marca colectivas* have also been used to protect certain Oaxacan handicrafts, notably woven textiles from Santo Tomás Jalieza and woodcarvings from San Martín Tilcajete, in the case of the latter from copies produced in workshops in China (Cant 2012). Caldo de piedra does not currently possess its own *marca colectiva* and the fact that it is a prepared dish rather than a single food product such as *cotija* cheese or a *tlayuda* may make it difficult to secure one.

Nevertheless, all of this points to an evolving regulatory apparatus and heightened appeals to the singular, proprietary nature of foods and foodways in Mexico. Not only are these foods rooted in particular geographies but they are often discursively situated within an ancient Mesoamerican past—however real or imagined. Oaxaca continues to self-consciously evolve into a center of history and culture for the global consumption. Both UNESCO designations and *marcas colectivas* act as powerful signifiers, drawing heightened attention to the things that they mark, conferring new social and economic value on food practices and products that may have otherwise gone unnoticed in Oaxaca's high profile tourism market.

The institutionalization of food as heritage and/or cultural property has opened up the floodgates to a host of new culinary claims. As a result, Oaxaca, like other world regions attracting international attention for their gastronomic heritage, not surprisingly is witnessing a proliferation of highly localized assertions of proprietorship over specific culinary products and practices. As Bourdieu (1993) famously theorized for artistic works, the meaning, value, and even the work itself is not produced from a top-down or even bottom-up process, but rather through contestation, positioning and position-taking among all actors involved within the field of production. Bourdieu's frame of analysis easily lends itself to understanding the various planes or levels in which food-based heritage claims and products are constructed and elaborated on. National and international agencies and organizations may create institutional frameworks through which material and other, less tangible forms of "culture" are made salient and are deemed worthy of preservation; UNESCO is perhaps the most famous international arbiter of heritage, but, as illustrated by Mexico's *marcas collectivas*—a system found not only in that country but throughout much of Latin America—the nation-state is also a critical player in the marking and legal protection of selective heritage and other cultural goods. But as the caldo de piedra case study illustrates, they ways in which individuals or groups on the ground interpret or operationalize officially sanctioned notions of heritage may deviate from the intentions of the over-arching state or international apparatus—and may even conflict with one another at the local level.

References

Benz, B. 2005. Archaeological evidence of teosinte domestication from Guilá Naquitz, Oaxaca. *Proceedings of the National Academy of Sciences* 98(4): 2104–6.

Bourdieu, P. 1993. *The Field of Cultural Production*. New York: Columbia University Press.

Breglia, L.C. 2006. *Monumental Ambivalence: The Politics of Heritage*. Austin: University of Texas Press.

Brulotte, R.L. 2012. *Between Art and Artifact: Archaeological Replicas and Cultural Production in Oaxaca, Mexico*. Austin: University of Texas Press.

Cant, A. 2012. *Practising aesthetics: artisanal production and politics in a woodcarving village in Oaxaca, Mexico*. PhD thesis, The London School of Economics and Political Science (LSE).

Di Giovine, M.A. 2009. *The Heritage-scape: UNESCO, World Heritage, and Tourism*. Lanham, MD: Lexington Books.

Haines, H.R. and Sammells, C.A. 2010. *Adventures in Eating: Anthropological Experiences in Dining from Around the World*. Boulder: University of Colorado Press.

Heldke, L.M. 2003. *Exotic Appetites: Ruminations of a Food Adventurer*. New York: Routledge.

Hobsbawm, E. and T. Ranger. 2003. *The Invention of Tradition*. New York: Canto.

Kennedy, D. 2010. *Oaxaca al Gusto: An Infinite Gastronomy*. Austin: University of Texas Press.

Kirshenblatt-Gimblett, B. 1995. Theorizing Heritage. *Ethnomusicology*, 39(3): 368–80.

____. 2004. Intangible Heritage as Metacultural Production. *Museum International*, 56 (1–2): 52–65.

Long, L.M. 2004. Introduction, in *Culinary Tourism*, edited by Lucy M. Long. Louisville: University of Kentucky Press, 1–19.

Martínez, Z. 1997. *The Food and Life of Oaxaca: Traditional Recipes from Mexico's Heart*. New York: MacMillan.

Mazatán-Páramo, Ricardo. 2010. *A Nation's Update: the Mexican Food Narratives and Cultural Diversity: Gastronomic Communication and the Intangible Cultural Heritage at the Juncture of Cosmopolitanism, Nationalism and Indigeneity*. Saarbrucken, Germany: VCM Verlag.

Pilcher, J.M. 1998. *¡Que vivan los tamales! Food and the Making of Mexican Identity*. Albuquerque: University of New Mexico Press.

Trilling, S. 1999. *Seasons of My Heart: A Culinary Journey Through Oaxaca, Mexico*. New York: Ballentine.

UNESCO 2010. Traditional Mexican cuisine—ancestral, ongoing community culture, the Michoacán paradigm: Nomination File no. 00400 for Inscription on the Representative List of the Intangible Cultural Heritage in 2010 [English] [Online: United Nations Educational, Scientific and Cultural Organization]. Available at: http://www.unesco.org/culture/ich/index.php?lg=en&pg=00011&RL=00400 (Accessed 11 August 2012).

Chapter 8

Hallucinating the Slovenian Way: The Myth of Salamander Brandy, an Indigenous Slovenian Psychedelic Drug

Miha Kozorog

Introduction

Let me begin with an ethnographic anecdote. In anticipation of the band NOFX at the Punk Rock Holiday festival in Tolmin, a small town in western Slovenia, which I attended as a festival researcher on August 15, 2011, I joined a group of people consisting of locals and visitors from the southern parts of Slovenia. One of the latter lit a joint and passed it around, saying that this part of Slovenia has a reputation as a place where everyone smokes marijuana, by which he was probably referring to "youth."[1] "But," he continued, "very near here, in the neighboring valleys, you also have salamander brandy, which is hallucinogenic." Somebody commented that, alas, he does not know of anybody making it. This was followed by the first man stepping into the role of expert:

> There are three ways of making it. The first is to throw a couple of salamanders into a barrel with soaked fruit so that while crawling around in the slush they release their poison, after which you use the fruit to make brandy. The second is to soak a salamander in brandy so that it releases its poison out of fear. The third way, which is also the most effective, is to hang a salamander under the pipe while you are making brandy so that the brandy trickles over the salamander and brings out the poison from its skin.

I inquired if they also made this drink where he is from, whereupon he indicated with his hand, "the home of salamander brandy is over there, behind those hills."

---

1 All those present were older than 35. "Youth" is generally defined as a transitive period in life, extending between childhood as a period of complete dependence on adults and adulthood as a period of existential independence from the family of origin. A different approach points toward "individualistic life-styles" as typical markers of youth. Thus, youth is no longer understood primarily as a temporal phase, but as increasingly becoming a social phenomenon of relevance to all generations (Ule and Miheljak 1995). In this chapter, too, the concept of youth refers mostly to this second meaning.

I asked him if he had ever tried this brandy. "No, but I know someone who could get it," he said, concluding our conversation, which had followed a path I was already familiar with.

For youth immersed in various forms of popular culture, drugs are part of the everyday, at least as a narrative (see, for example, Manning 2007, Rubin 2006, Sanders 2006). For instance, popular music is full of references to them. In addition, mass media have made drugs a common theme in everyday news (South 1999: 7–8). However, in the mid 1990s the story of an "autochthonous Slovenian hallucinogenic drug" broke out of the usual media routine and became an especially popular topic of conversation among youth in Slovenia. Salamander brandy was attractive because it was fresh, exotic and "Ours." When on March 4, 2005, again in Tolmin, I was waiting for a concert by Options, a jazz band from New York, I overheard the band members' conversation with a local youth. The latter informed them that we Slovenians have our own autochthonous hallucinogen, which the visitors acknowledged as an interesting curiosity. Thus, for at least a particular segment of Slovenians, salamander brandy became a marker of identity, a part of material culture that they could use to proudly represent themselves within contemporary pop-cultural flows.

A "currency" such as salamander brandy can have a global value,[2] at least for people with psychedelic affinities, as is demonstrated by an American looking to contact someone from Slovenia on the Internet who could help him open "the doors of perception":

> So, I'm a little bit of a psychonaut and I enjoy trying new things. Recently I picked up a novel and read about a spirit called Salamander Brandy. ... People online have said by talking to people in towns it can be found and that it sells for around 15$ a bottle (meaning a nice little profit for whomever secures them for me as I would be buying several bottles, and possibly more in the future). (Request 2011)

However, this particular material culture only exists in stories. Slovenians can mention salamander brandy to make an impression on foreigners, but we cannot pour them a glass or send them a bottle. The problem is deeper than it appears at first glance. It is not the brandy's hallucinogenic properties and connection to drug culture—both of which it symbolically represents and legitimizes—that is problematic, but that the drink is practically non-existent. When trying to procure it, one will always find somebody who knows somebody, and so forth. Moreover, one also finds that almost no one has ever tried the drug, although there do exist the very few who by following

---

2 The use of economic terminology is not a coincidence. Building upon Bourdieu's notion of "cultural capital," Sarah Thornton (1995) introduced the concept of "subcultural capital," by which she refers to knowledge and experience that is effective only within particular (youth) groups and confers upon individuals unequal statuses within such groups.

recipes (see introductory paragraph) have produced the drink themselves. Nevertheless, salamander brandy does function (at least on a verbal level) as a *representation* (Hall 2003), and therefore points to the essence of heritage, which lies foremost in its symbolic (rather than material) character. Heritage "is not so much a 'thing' as a set of values and meanings" (Smith 2006: 11). Although salamander brandy is materially non-existent, it is "good to think with" (cf. Muršič 2005), good for establishing symbolic relations.

When thinking through heritage, one must therefore also think about the relations that give a heritage substance. One must take into account the cultural frameworks that grant something of "Ours" representative status. Therefore, besides a thing being "Ours" (and old), the cultural contexts in which it functions (here and now) are also of importance. The contexts provided by tourism (see, for example, AlSayyad 2001) are undoubtedly different from those provided by the diverse field of popular culture. For salamander brandy, at least two interrelated cultural frameworks are relevant: the popular drug culture with roots in the 1960s, and New Age movements from the end of the millennium, which in western societies, Slovenia being no exception (Lesjak 2001), as a diffuse field of practices related to spirituality and personal growth spread awareness about indigenous practices of self-exploration. Although hallucinogenic drugs are ostensibly not an important part of these practices, they are nevertheless connected to New Age (Chryssides 2007: 6, Hanegraaff 1996: 11, 50–51, Partridge 2005). Through salamander brandy, "Slovenian culture" therefore becomes affiliated with cultures that count as "cool" in this interwoven imaginary of hallucinogenic drugs.

Despite these affirmative contextualizations, the national heritaization of salamander brandy is not a simple matter. Drugs are tabooed by state policy. In addition, its production apparently involves the use of live salamanders (*Salamandra Salamandra*) and therefore constitutes animal torture. The heritagization of salamander brandy has thus mainly been left to grass-roots actors fascinated by the exotic. According to my fieldwork investigations, however, salamander brandy is not really exotic. The label "salamander brandy" was used for hard liquor of an inferior quality. It refers to practices of home alcohol distillation that could never be constructed into any kind of representational symbol, above all not as part of a "hallucinogenic heritage."

A Sensational Discovery

My introductory anecdote shows that some individuals possess detailed knowledge of salamander brandy. In similar conversations one often encounters an "expert" who knows the three procedures for making the drink and its geographic origins. The source of this expert discourse can be located precisely. In 1995 Blaž Ogorevc, a well-known Slovenian writer, journalist, traveler and representative of the 1960s hippie generation, published an article in the weekly

Mladina titled "Salamander Brandy: Hallucinogenic Drugs—Made in Slovenia," which contained the following message: alongside typical home-made liquors, traditional Slovenian peasant culture also produced special hallucinogenic ones. To the national heritage of peasant alcohol distillation was thus added an exotic product comparable to "mind-expanding" drugs in other cultures. The messages and descriptions of this original article have become part of folklore.

To explain the article's impact, several levels of communication must be taken into account. Let us begin with the author, who as mentioned belongs to the 1960s hippie generation, which is considered to be the cradle of the psychedelic drug culture. Salamander brandy also originates in the part of Slovenia where he lives. As a personification of the 1960s and a local, readers find him trustworthy.

The article is written in a humorous literary style that leaves much room for interpretation, but behind the literary wrapping the reader is served facts. Ogorevc discovered salamander brandy by finding a person making it and tested its effects himself—exemplary investigative journalism, therefore. Since he found it "in the field" and confirmed its effects through "participant observation," the reader is compelled to believe him. Moreover, the described hallucinogenic experience, despite literary exaggeration, was exactly like those familiar from literature: hallucinations, sexual stimulation, and objects radiating on their edges.

The author was prompted to research salamander brandy by a film scene from San Francisco,

> ... a city that for decades has been dictating to western culture the fashion of taking drugs ..., in which some aged drug freak ... now and then ... picks up a fat Mexican toad and delightfully licks its warty back . ... While they were all puffed up on account of that San Francisco of theirs ..., my heart felt the sting of local-patriotism. We, here in [Škofja] Loka [Ogorevc's home place], have in comparison something so beautiful and elegant, salamander brandy .... And not as a fashionable fad ..., but [as] an elixir ennobled and strengthened through centuries-long tradition. (Ogorevc 1995: 27)

The text crafts therefore salamander brandy as an autochthonous and authentic product. It also mentions a Chinese tradition of soaking another species of amphibian in brandy (1995: 30) and includes the assumption that "every country, every region, has its drug" (1995: 29). Thus the author emphasizes the cultural background of Slovenian psychedelic practice, which like those of other cultures has its roots in the distant past. According to the author, the practice was invented by witches (1995: 29) and is therefore at least some centuries old. The phenomenon is thus unambiguously located in time and place: from an unknown past it has been preserved among people in Slovenia to this day, when the researcher discovered and described it. The drink is cultural heritage *par excellence*.

A Hallucinogenic Heritage

For something to become heritage the process of heritagization is necessary. This entails the construction of a special symbolic value for a particular tradition (or antique object, landscape, etc.), which thus becomes a symbolic representative of a particular community. After 1991, when Slovenia seceded from Yugoslavia and became a democracy, the process of heritagization was also democratized, diversified and liberalized. Heritage thus increasingly became not only of national and local but also of entrepreneurial interest. In addition, because of its role in the struggle for Slovenia's independence, popular culture, such as punk (Muršič 2005: 25), became part of national heritage as well. It was in this context of youthfulness and optimism characteristic of the birth of a new state that in 1995 Ogorevc introduced the Slovenian public to salamander brandy, thus contributing a provocative deviation from Slovenia's alcohol norm.

Heritagization is authorized by those possessing specific knowledge conferred upon them by the state, which is the result of a particular historical process (see Smith 2006: 17–28). Despite this, traditions and antiques can also occupy representative positions by by-passing designated experts (cf. Davison 2008, Smith 2006). One would hardly expect these experts to be involved in the field of drug culture, and so Ogorevc became a suitable authority who could position salamander brandy among national representational symbols, at least for some Slovenians. For them it became national and not merely local heritage. Although it comes from a specific area, already in the title of the article Ogorevc also attributes to it a national value, while in the text he places salamander brandy alongside the then-president of Slovenia, Milan Kučan, and *Proteus anguinus,* an animal species endemic to Slovenia (1995: 31). However, to understand why for individuals, especially youth, precisely a hallucinogenic drink was suitable as a representational symbol, it is necessary to understand the global cultural flows that legitimize it. As we have seen, Ogorevc found support for this tradition of "Ours" in western and indigenous drug cultures.

In western societies, drugs have a numinous character since they are simultaneously terrifying (*mysterium tremendum*) and attractive (*mysterium fascinans*). Although by themselves they perhaps indeed possess these attributes, as here deduced from Otto's analysis (1970 [1917]) of the sacred, which is why they are encountered in numerous religious traditions (see, e.g., Harner 1973), in western societies they are also present through a constant repetition of particular notions and images. Drugs are an enemy with whom the USA is at war, an unconditional evil leaving "victims" in its wake. On the other hand, drugs appear in the name of one of the most globally established brands, Coca Cola (which is also a U.S. export).

However, western popular culture has at least since the 1960s also perpetuated an imaginary according to which drugs—albeit potentially dangerous (popular culture harbors a collective memory of overdoses by its stars)—are a fuel without which the inspiration for the timeless creations of this culture would supposedly not exist (Partridge 2005: 99). Especially "mind-expanding" drugs, with marijuana and hallucinogens at the forefront, are seen as belonging in this category. Thus Timothy Leary, "the high priest of LSD," cited authors such as James Joyce, Alan Watts and beatniks Allen Ginsberg and William Burroughs as precursors or fellow travelers of his psychedelic movement (Rudgley 1993: 106). This popular drug culture, which exploded in the 1960s, has deepened the memory of this creative side of drugs to include not only beatniks and the research of Aldous Huxley (1994 [1954]) but also Charles Baudelaire in the nineteenth century (his poem *Les Paradis Artificiels* from 1860 is about hashish and wine) and "centuries-long" indigenous traditions from across the globe.

Furthermore, psychedelics and indigenous practices of self-exploration are also connected to a new spirituality. At the end of the 1960s anthropology student Carlos Castaneda carved a deep groove into the imaginary of self-exploration with his research on mescaline-based spiritual practices of indigenous Yaqui people in Mexico. His works, despite being discredited by anthropologists (see, e.g., Beals 1978),[3] remained an important source of New Age inspiration throughout the 1990s. Atkinson thus finds that the new shamanism in the United States and Europe was "spawned by the drug culture of the 1960s and 1970s, the human potential movement, environmentalism, interests in non-western religions, and by popular anthropology, especially the Castaneda books" (1992: 322). Just in 1995, the year Slovenia "obtained" salamander brandy, as many as eight books by Castaneda were published in Slovenian as part of a single collection. From this point of view as well, the Slovenian hallucinogen appeared at the right time.

In the context of New Age spirituality, indigenousness is an important aspect in the evaluation of a drug. In her analysis of the BBC's television series *Tribe,* Pat Caplan finds that in constructing tribalism its authors also referenced indigenous hallucinogenic drugs, and identified people interested in New Age philosophies as a typical audience of the series (2003). De Rios and Rumrrill write about New Age tourism based on the experience of ayahuasca: "The Amazon is said to be the last remaining spiritual sanctuary on earth. By paying the cost of the trip, one becomes an impeccable warrior in the tradition of the writer Carlos Castaneda" (2008: 70). Benjamin Feinberg sheds light on the connection between hallucinogenic mushrooms, the late mushroom priestess María Sabina and New Age drug tourism in Mexico (2003). Ascribing indigenousness to a drug implies its archaic origins and thus also the "culturedness" of its usage:

> Highly toxic mescal beans, psilocybin mushrooms, lysergic acid diethylamide (LSD)-containing seeds of the morning glory plant, and the mescaline-containing tips of the peyote cactus were used ritually in the Americas thousands of years before the birth of Christ. (Robson 1994: 69)

---

3   Castaneda initially succeeded in convincing some anthropologists (e.g. Harner 1973: 6).

The controversial area of drug use thus moves into the area of cultural values, and experts (such as the one cited in the paragraph above) can then also contribute an affirmative discourse (see, for example, Ott 1997). Such discourse, as I will show, was gradually also provided for salamander brandy.

Between Media and Science

Drugs are not easily rendered as a form of heritage. Thus, for the heritagization of salamander brandy an actor was necessary that would have no problem with controversy. Such was, besides Ogorevc himself, the weekly *Mladina*, serious and trustworthy on the one hand and subversive and provocative on the other. The weekly enjoys a cult status in Slovenia. During the 1980s it was seen across Yugoslavia as the publication most uncompromising in revealing the nature of the regime. It is therefore a politically engaged publication, not only in a narrow political sense but also in the sense of initiating public debate and shaping Slovenian society in general. It has frequently attacked social taboos, published contributions on subcultures and addressed issues relating to minorities, sexuality, and also drugs.

Ogorevc's article in *Mladina* is eight pages long. Because of its length, style and message, it became practically anthological. *Mladina* also took up the promotion of this discovery and occasionally reminded readers of it. In an article on the drug trade, Urbaničič (1997: 10) included the anecdote about salamander brandy. The brandy then also appears, completely out of context, in connection to Slovenia's voting system: "Before voting on electoral districts politicians should drink a bottle of salamander brandy so that the division of Slovenia would finally be settled. After imbibing a liter of salamander brandy, it would surely be easy to make political compromises, and the majoritarian voting system would be affirmed" (Trampuš 2000: 7). This was followed by an article that echoes the context of Ogorevc's discovery of unusual brandy making, although it does not directly refer to salamander brandy (Talibečirović 2003). A more direct reference to the topic can be found in a feature on Vietnam:

> I don't know what drove me to eat snake meat. It was probably pure investigative enthusiasm, inspired by the example of legendary Slovenian investigative journalist Blaž Ogorevc. This holds true especially for trying various hard liquors in which different kinds of snakes had been soaked. ... While in the Polhograjski Dolomites they brew a drink whose main characteristic is provided by the glands of an amphibian, in the abundance of drinks accompanying a snake feast one may encounter numerous effects similar to those of the famous salamander brandy. (Kocjan 2008: 49)

Mladina therefore actively maintained salamander brandy in the consciousness of its readers. It also made a lasting impression with a photo of Denis Sarkić featuring a bottle filled with green liquid with a salamander floating in it and sealed with blue wax (in Trampuš 2000: 7).[4] Thus salamander brandy also obtained a visual dimension. However, I am quite convinced that the bottle on the photo was not a product of the society in which Ogorevc "discovered" salamander brandy but was fabricated, since one does not find waxed bottles in peasant alcohol production, although one does find green liquors (in which herbs are soaked), as is also the case in the photo.

During this time salamander brandy was also noticed by other researchers. In 1998 Ivan Valenčič, an amateur researcher of hallucinogens, published the article "Salamander Brandy: A Psychedelic Drink Made in Slovenia" in the German *Yearbook for Ethnomedicine and the Study of Consciousness*. Valenčič apparently drew on informers who had been in contact with salamander brandy. As his writing reveals, however, these "informers" were primarily taken from Ogorevc's article. At times his text verges on plagiarizing Ogorevc: "If the brandy is of good quality then it must elicit not only pleasant inebriation but also all those things that had been seen by Master Hieronymus Bosch ..." (Valenčič 1998: 220); Ogorevc writes that drinking the brandy "causes in its users not only the drunkenness encountered at a firemen's ball but also the things seen by Master Hieronymus Bosch" (1995: 31). Valenčič cites Ogorevc selectively, at times by name while elsewhere presenting his words as those of others—in the case of Bosch's visions he puts words in the mouth of "an experienced person" (1998: 220).

The difference between the two articles lies primarily in that Valenčič wrote unambiguously, without humor and literary additions, therefore scientifically. He does give some new information, for example the rich symbolism of the salamander, which he traces back to the Middle Ages (1998: 213, 215, 216, 217, 223), but symbolic genealogy does not validate the conclusion that the practice of making salamander brandy originated in the Middle Ages (1998: 213), especially since the distillation of hard liquor appeared in Europe only in the fifteenth and especially the seventeenth century (Montanari 1994: 123–4), while home distillation was probably not present in the area in question until the nineteenth century (Košir 2009: 30, Sušnik 1958: 176).

Valenčič's claim about the effects of salamander poison is also problematic (1998: 221–3). Although he uses chemical formulas to prove hallucinogenic properties, he concludes that scientists have not yet done explicit research on the psychoactive properties of salamander toxins (1998: 224). Even more surprising are some of his ethnographic descriptions. The article begins with the following:

> In the hilly region northwest of the Slovenian capital of Ljubljana, every autumn a very unusual event takes place: a salamander hunt. There are many salamanders in the rainy forest hollows and in some places the hunters must take care not to step on them. (Valenčič 1998: 213)

---

4   I have tried to obtain information on this photo from its author without success.

As yet this hunt has only been observed by Valenčič, although his narrative gives the impression of a collective practice. Also new is the fact that wormwood is added to salamander brandy to make it tastier and to enhance its psychedelic effects (1998: 200). Given that Valenčič did not observe the making of the drink, one can only speculate where he got these ethnographic details.

However, the discourses that continued the construction of salamander brandy were not always positive, as in *Mladina* or Valenčič's article, but also drew on the fear of drugs and allegations of animal torture, which is how the sensationalist POP TV chimed in. Reporter Ana Jud's contribution easily reads as a television adaptation of Ogorevc's writing. The *Mladina* article features photos of a hilly, misty forest landscape, and this "mystical atmosphere" is also reproduced in the television version. Over this visual backdrop the already familiar story about a Slovenian hallucinogenic drug is presented. The reporter cites Valenčič as an authority who, despite not wishing to appear on camera, compares the effects of salamander brandy to those of LSD. The cover of a publication by Valenčič is also shown, featuring the photo of Sarkić from *Mladina*. However, there is an important difference between the TV report and the *Mladina* article. Whereas the latter contributed to a positive attitude toward salamander brandy, POP TV also evoked fear by raising concerns about animal torture and substance abuse among youth. In the report a young man recounts the three recipes for making salamander brandy and describes its effects in Ogorevc's style as extremely sexually stimulating. The report concludes with the moral that salamander brandy used to be made by old people, while today it is made by young men, but that in the past it was acceptable, "since the ancient ingredients were picked exclusively from nature," while today all that matters is the potency. Thus the TV report achieved its aim: the Slovenian public was given the chance to feel shocked, while the sensation of a Slovenian hallucinogen was given publicity. In spite of this sensationalist "problematization" of salamander brandy, a positive image prevailed, at least among youth.

As proof of authenticity, Valenčič cited the chemical formulas of salamander toxins. Irena Maček Ložar, a graduate student of pharmacology, also approached the topic from this angle. Her intent was to "[describe] in more detail the toxins of the common salamander … and salamander brandy" (Maček Ložar 2003: 719). However, her article offers no empirical evidence that salamander toxins cause hallucinations. "Some sources claim that the dermal extracts of salamanders contain signaling compounds for intra- and interspecific recognition, but the precise nature and source of these substances have not been established" (2003: 724). She thus references Ogorevc's descriptions of the hallucinogenic properties of the toxins. She also looks for parallels in other cultures: "Preparations made from amphibians are consumed across the globe. In the U.S. getting high with toad poison is popular. … Some preparations of Chinese origin contain a high level of bufotoxin. … Some South American natives …" and so on (2003: 724). The article concludes: "[W]here salamander brandy has resided for centuries, there still circulate many true stories and legends about this intoxicating drink, while some make and drink it in secrecy even today" (2003: 725).

However, salamander brandy did not attract attention exclusively in Slovenia but also abroad, among both the lay public (see, e.g., Kick 2004: 45) and scholars (see Blom 2010: 227–8).[5] An internet article by John Morris (n. d.) with the photo of Sarkić inspired several internet posts and forum debates outside Slovenia. Morris traveled across Slovenia as a guest of the Slovenia Tourist Board, but he gives special thanks to Ogorevc, "an authority on all things strange in Slovenia." His article, subtitled "A medieval method of getting in touch with your deeper sexual feelings," features the already familiar portraits, but the reader is also informed that the drink can be procured:

> The Slovenian tourist board certainly has not heard of it. But the secrecy of the brandy is half the fun. I had to go through a chain of whispered contacts and endless hours in smoke-filled taverns in Škofja Loka before finding the stuff. (n.d.)

One should also mention Slovenian ethnologists, who had already been conducting research, by disciplinary tradition, on peasant culture in Slovenia. Ethnologists have documented dietary and brandy-making traditions, including those in the area of Škofja Loka (Sterle 1987, Sušnik 1958), but have written nothing about salamander brandy. However, when the *Slovenian Ethnological Lexicon* was published in 2004, salamander brandy also found a place in it. In this case, too, Ogorevc's *Mladina* article served as the primary reference:

> In the area around Škofja Loka salamander brandy is known, which men use as an—*aphrodisiac*. A live salamander is put in the fruit used to make brandy, or into a bottle of brandy, or is tied to the pipe of the distillation barrel so that it releases toxic hallucinogenic substances. (SEL 2004: 332)

Salamander brandy therefore aroused the interest of various publics, while its main characteristics remained as defined by Ogorevc, despite differing interpretations. These characteristics made Slovenians more exotic and interesting (for others and for themselves). They were mostly identified with by youth, who saw in the drink a suitable representative of national traditions and their own (national) identity. Primarily in their case one can talk of heritage as representation, which of course is not something displayed in museums or promoted by the tourism industry, but something alternative to all this and at the same time in dialogue with the popular heritages of the world.

---

5   Both Kick (2004) and Blom (2010) also referred to my article (Kozorog 2003) as contradicting the common portrait of the drug.

An Impotent Legacy[6]

Salamander brandy functions as heritage because it draws the necessary aura from the global cultural frameworks of psychedelic drugs. Without this psychedelic aura, we would be left with only the heritage of home brandy making, which by itself is a notable part of national heritage (see Šertel 1989). However, my thesis (see also Kozorog 2003) is that salamander brandy in fact lacks both auras—the one provided by psychedelia as well as the one that puts home brandy on the pedestal. It is exactly the opposite of the latter, namely a distortion of notions about "pure home brandy making." In traditional society, "salamander brandy" was a label for "bad brandy," for fraudulent and inferior-quality alcohol distillation. As such it is interesting neither in the context of drugs nor in the context of home-made brandy. It is merely an impotent legacy and would be very difficult to elevate to the status of heritage.

In his pivotal article Ogorevc describes how he obtained the mysterious drink. When drinking brandy at some farmer's place, he convinced the farmer to also serve and sell him a special brandy—salamander brandy. However, he came to suspect that the farmer had not really served him a brandy different from the one he was already drinking, although he had paid more for it. He therefore concluded that "one always has to pay for ideology" (1995: 30). Thus, Ogorevc set out with an idea of what he was looking for and so missed a crucial fact. As he found out, adding salamanders makes the brandy "flow more" (1995: 30). Informal expressions such as "flows more" and "gives more" are used by Slovenian brandymakers to refer to the quantity of brandy produced from a given amount of ingredients. The journalist Jud (2002) was also told by a farmer that salamander brandy "flowed immensely." Such expressions are not new. When in the second half of the nineteenth-century doctors talked of a brandy-drinking epidemic, in 1884 one of them pointed to an inferior brandy called "fuzel" in particular, which was made so that the procedure "gives more" (Studen 2009: 77). I, too, was informed that adding salamanders to fruit "gives more" of the brandy while I was researching it in the area around Škofja Loka.

Increased quantities were achieved in traditional society in various ways, be it by adding bought spirits or by brewing with potatoes, sugar beet, wheat, maize, wild chestnuts and forest berries, or simply by brewing faster and longer (personal field notes, Košir 2010: 30, Slomšek 1847: 22, Sterle 1987: 119, Studen 2009: 42, 73, 77, Sušnik 1958: 168–9), since a good brandy is brewed slowly and only until the alcohol content remains above a certain level. The reason was economic. Salamander brandy belongs to these inferior brandies produced in larger quantities, as good brandy is made from fruit (apples, pears, plums etc.). Whether with the addition of salamanders brandy really "just flows" is not known. I leave open the possibility that with its yellow

---

6   While by "heritage" I refer to that which is a source of pride and a support to identity, by "legacy" I refer to the value-neutral effects of the past on the present, that is, all that is carried over from the past and leaves a trace in our world (see Baskar 2005: 46–7).

spots the salamander simply evoked a feeling of danger and uneasiness, and therefore became symbolically linked also to dangerous and impure brandy (cf. Douglas 2002 [1966]).

In the area of Škofja Loka no one mentions salamander brandy in connection to altered states of consciousness, but exclusively in connection to fraudulent brandy making. The term was used when after drinking brandy very negative effects were felt, such as partial paralysis and sickness, or when a farm produced more brandy than it could have given its available resources. People never ordered salamander brandy, but bought or drank it unwittingly. It was sold by fraudsters or people from the margins of society for whom quantitatively-oriented brandy making was a means to achieve the existential minimum.

"We have always been honest brandy makers" and "making salamander brandy is a shameful act" are characteristic statements by my informants during my fieldwork. Because fraud with inferior brandy is shameful, no one wants to be associated with it. However, a folktale did connect salamander brandy to a concrete woman, who passed away in 1952, leading me to visit the farm on which she had lived. The farm is located on the periphery, geographically as well as symbolically. From my conversations with the woman's son and daughter-in-law as well as other locals, I was not able to determine the veracity of the rumors that she had brewed salamander brandy. But from what I was told it seems plausible that she had simply been stigmatized for not conforming to the social norms of traditional society, and that the folktale functioned as a public barb against her. She and her husband had been poor, with nine children to support, and the allegations connecting her to salamander brandy seem to had started as punishment for organizing on the farm a "man's world," including socializing, drinking alcohol, gossiping, and probably also eroticism, by allowing men, including guest workers from other parts of Yugoslavia, to gather there and also bring local girls. Salamander brandy thus served as a handy means of social discreditation.

The world of hallucinations is thus behind us, and we have entered the real world of social sanctions and marginalization in rural Slovenian society. Whether people also used salamanders to increase brandy production, as some have said to me, is not known for sure. What is sure is that those who drank "bad brandy" and felt its ill effects, or those who wanted to publicly discredit another person, did not require proof in the form of salamanders. It was enough to feel that "something is not right."

And it is for this same reason, that "something is not right," that salamander brandy appears as problematic for heritagization. First of all, with its reference to drugs it is perhaps not the "right" drink to easily accept into expert heritage discourse. However, after being presented as an indigenous hallucinogenic drug by actors who possessed the cultural capital necessary to legitimize it within certain globally established (popular) cultural frameworks, salamander brandy might indeed also become a part of expert heritage discourse—as in the case of the *Slovenian Ethnological Lexicon*. Since drugs are not only *tremendum* but also *fascinans*, they might in our "youthful" and "joyous" society to a certain extent gain acceptance and thus provide a specific popular connotation to (national) heritage. However, the "rightness," "suitability" that is the issue here does not refer to the social status of drugs but to the social status of those who produced "real" salamander brandy. Or to put it differently, granting legitimacy to the invented or imagined version of salamander brandy, with its references to hallucinations and so also to widely recognized popular culture, would not be as problematic as granting legitimacy to the "real" salamander brandy, which is a product of social marginalization. The real history of salamander brandy points to troublesome aspects of traditional rural Slovenian society, and it is this that appears as an obstacle to making salamander brandy a national heritage symbol.

References

AlSayyad, N. 2001. Global Norms and Urban Forms in the Age of Tourism: Manufacturing Heritage, Consuming Tradition, in *Consuming Tradition, Manufacturing Heritage: Global Norms and Urban Forms in the Age of Tourism*, edited by N. AlSayyad. London and New York: Routledge, 1–33.

Atkinson, J.M. 1992. Shamanisms Today. *Annual Review of Anthropology*, 21, 307–30.

Baskar, B. 2005. Avstro-ogrska zapuščina: Ali je možna nacionalna dediščina multinacionalnega imperija? [Austro-Hungarian Legacy: Is National Heritage Based on a Multinational Empire Possible?], in *Dediščina v očeh znanosti*, edited by J. Hudales and N. Visočnik. Ljubljana: Filozofska fakulteta, 41–52.

Beals, R.L. 1978. Sonoran Fantasy or Coming of Age? *American Anthropologist*, 80(2), 355–62.

Blom, J.D. 2010. *A Dictionary of Hallucinations*. New York: Springer.

Caplan, P. 2005. In Search of the Exotic: A Discussion of the BBC2 Series "Tribe.". *Anthropology Today*, 21(2), 3–7.

Chryssides, G.D. 2007. Defining the New Age, in *Handbook of New Age*, edited by D. Kemp and J.R. Lewis. Leiden and Boston: Brill, 5–24.

Davison, G. 2008. Heritage: From Patrimony to Pastiche, in *The Heritage Reader*, edited by G. Fairclough et al. London and New York: Routledge, 31–41.

De Rios, M.D. and Rumrrill R. 2008. *A Hallucinogenic Tea, Laced with Controversy: Ayahuasca in the Amazon and the United States*. Westport and London: Greenwood.

Douglas, M. 2002 [1966]. *Purity and Danger: An Analysis of the Concepts of Pollution and Taboo*. London and New York: Routledge.

Feinberg, B. 2003. *The Devil's Book of Culture: History, Mushrooms, and Caves in Southern Mexico*. Austin: University of Texas Press.

Hall, S. 2003. Introduction, in *Representation: Cultural Representations and Signifying Practices*, edited by S. Hall. London, Thousand Oaks and New Delhi: Sage, Milton Keynes: The Open University, 1–11.

Hanegraaff, W.J. 1996. *New Age Religion and Western Culture: Esotericism in the Mirror of Secular Thought*. Leiden, New York and Köln: Brill.

Harner, M.J. (ed.) 1973. *Hallucinogens and Shamanism*. London, Oxford and New York: Oxford University Press.

Huxley, A. 1994 [1954]. *The Doors of Perception and Heaven and Hell*. London: Flamingo.

Jud, Ana, 2002. Preverjeno. [Verified.] *POP TV*, 21 May.

Kick, R. 2004. *Book of Lists: Subversive Facts and Hidden Information in Rapid-fire Form*. New York: The Disinformation Company Ltd.

Kocjan, A. 2008. Oprosti, ker sem te pojedel, čudovita kača! [Sorry for Eating You, Wonderful Snake!] *Mladina*, 33, 48–53.

Košir, T. 2010. Življenje na Lučinskem skozi stoletja. [*Life in Lučinsko throughout the Centuries*.] Lučine: Krajevna skupnost.

Kozorog, M. 2003. Salamander Brandy: A "Psychedelic Drink" between Media Myth and Practice of Home Alcohol Distillation in Slovenia. *The Anthropology of East Europe Review*, 21(1), 63–71.

Lesjak, G. 2001. Nova religijska in duhovna gibanja v Sloveniji. [New Religious and Spiritual Movements in Slovenia.] *Teorija in praksa*, 38(6), 1108–24.

Maček Ložar, I. 2003. Strupi dvoživk in močeradovec. [Amphibian Toxins and the Salamander Brandy.] *Farmacevtski vestnik*, 54, 719–25.

Manning, P. (ed.) 2007. Drugs and Popular Culture: Drugs, Media and Identity in Contemporary Society. Portland: Willan Publishing.

Montanari, M. 1994. *The Culture of Food*. Oxford and Cambridge: Blackwell.

Morris, J. n.d. Salamander Brandy: A Medieval Method of Getting Touch with Your Deeper Sexual Feelings. Available at: http://www.grailtrail.ndo.co.uk/ Grails/brandy.html (Accessed 20 June 2011).

Muršič, R. 2005. Kvadratura kroga dediščine: Toposi ideologij na sečišču starega in novega ter tujega in domačega [Squaring the Circle of Heritage: Ideological Topoi at the Intersection of Old and New, and Foreign and Native], in: *Dediščina v očeh znanosti* (ed.) J. Hudales and N. Visočnik. Ljubljana: Filozofska fakulteta, 25–39.

Ogorevc, B. 1995. Močeradovec: Halucinogene droge—Made in Slovenia. [Salamander Brandy: Hallucinogenic Drugs—Made in Slovenia.] *Mladina*, 23, 26–32.

Ott, J. 1997. *Pharmacophilia: The Natural Paradises*. Kennewick: Natural Products Co.

Otto, R. 1970 [1917]. *The Idea of the Holy: An Inquiry into the Non-rational Factor in the Idea of the Divine and Its Relation to the Rational*. London, Oxford and New York: Oxford University Press.

Partridge, C. 2005. *The Re-enchantment of the West: Alternative Spiritualities, Sacralization, Popular Culture and Occulture, Volume 2*. London and New York: T&T Clark International.

Request. 2011. Request: I need to find someone living in Slovenia. Available at: http://www.reddit.com/r/Favors/comments/fansg/request_i_need_to_find_ someone_living_in_slovenia/ (Accessed 23 June 2011).

Robson, P. 1994. *Forbidden Drugs: Understanding Drugs and Why People Take Them.* Oxford, New York and Tokyo: Oxford University Press.

Rubin, L.C. (ed.) 2006. *Psychotropic Drugs and Popular Culture: Essays on Medicine, Mental Health, and the Media.* Jefferson: McFarland & Co. Inc.

Rudgley, R. 1993. *The Alchemy of Culture: Intoxicants in Society.* London: British Museum Press.

Sanders, B. (ed.) 2006. *Drugs, Clubs and Young People: Sociological and Public Health Perspectives.* Farnham: Ashgate.

SEL. 2004. *Slovenski etnološki leksikon.* [*Slovenian Ethnological Lexicon.*] Ljubljana: Mladinska knjiga.

Slomšek, A.M. 1847. Čujte, čujte, kaj žganje dela! *Prigodba žalostna ino vesela za Slovence.* [*Hear Ye! Hear Ye! What Brandy Does! A Sad and Happy Tale for Slovenians.*] Celovec.

Smith, L. 2006. *Uses of Heritage.* London and New York: Routledge.

South, N. 1999. Debating Drugs and Everyday Life: Normalisation, Prohibition and "Otherness," in *Drugs: Cultures, Controls and Everyday Life*, edited by N. South. London: Thousand Oaks and New Delhi: Sage, 1–15.

Sterle, M. 1987. Prehrana na Loškem. [Diet in Loško.] *Loški razgledi*, 34, 105–61.

Studen, A. 2009. *Pijane zverine: O moralni in patološki zgodovini alkoholizma na Slovenskem v dobi meščanstva.* [*Drunken Bests: On the Moral and Pathological History of Alcoholism in Slovenia during the Age of the Bourgeoisie.*] Celje: Zgodovinsko društvo Celje.

Sušnik, L. 1958. O žganjekuhi v Brezovici pod Lubnikom. [On Brandy Making in Brezovica pod Lubnikom.] *Slovenski etnograf*, 11, 167–78.

Šertel, A. 1989. *Domače žganje.* [*Home-Made Brandy.*] Ljubljana: Kmečki glas.

Talibečirović, A.L. 2003. Okus po zimi: Drekovec, žganje posebne vrste. [A Taste of Winter: Shit Brandy, A Special Kind of Brandy.] *Mladina*, 46, 50–51.

Thornton, S. 1995. *Club Cultures: Music, Media and Subcultural Capital.* Cambridge: Polity Press.

Trampuš, J. 2000. Politični interes in močeradovec. [Political Interests and Salamander Brandy.] *Mladina*, 14, 7.

Ule, M. and Miheljak V. 1995. *Pri(e)hodnost mladine.* [*The Transitivity and Future of Youth.*] Ljubljana: Državna založba Slovenije.

Urbančič, M. 1997. Trg z mamili: Cenik snovi, dostopnih na črnem trgu, ki uradno štejejo za prepovedana mamila. [The Drug Market: A Pricelist of Substances Available on the Black Market Officially Classified as Illegal Narcotics.] *Mladina*, 46, 10.

Valenčič, I. 1998. Salamander Brandy: A Psychedelic Drink Made in Slovenia. *Jahrbuch für Ethnomedizin und Bewußtseinsforschung* 1996–5, 213–25.

Chapter 9

Haute Traditional Cuisines:
How UNESCO's List of Intangible Heritage
Links the Cosmopolitan to the Local

Clare A. Sammells

Let me begin with a superficially self-evident statement: people do not eat the same ways at home as they do in restaurants. In domestic household kitchens, food is usually prepared by unpaid family members (or underpaid domestics), often women. These foods are prepared in set quantities for the members of the household, and respond to the particular lives of individuals who are intimately acquainted with each other. Such meals meet general cultural expectations, but also accommodate the individual desires of household members, as well as the constraints of particular work schedules and budgets. The same group of individuals participates (to varying degrees) in preparing, serving, and consuming food. (For example, see Counihan 1988, Counihan and Van Esterik 2007, Dunn 2004, Gill 1994, and Weismantel 1988, among many others.)

Restaurant kitchens, in contrast, rely on the paid labor of trained professionals. Food must be prepared in larger quantities, and respond to some individual requests, although not with the same level of intimacy as domestic meals. Serving food is often performed by different people from those either preparing or consuming the food. At least this is true in the public space of the restaurant: staff may also consume the food they sell, although usually not in the same spaces or contexts as paying clients. The public nature of restaurants is reflected in what people wear and how they act, from dressing up to go out to eat, to uniforms associated with restaurant staff and chefs. These economic and social relationships are very different from those found in a household kitchen, and extend into the realm of the food itself; ingredients, cooking techniques, and presentation are all interconnected with these social realities. If we accept the idea that foods are "good to think" as well as "good to eat" (following Levi-Strauss), it should not surprise us that household and restaurant meals differ in terms of both the food itself and the social interactions that surround cooking and eating in these spaces.

I have argued elsewhere that touristic cuisines are best understood as having their own culinary logic, rather than assuming that they are either attempts to faithfully imitate local domestic cuisine, or to recreate the foreign domestic foods that tourists might be familiar with from home (Sammells 2010). If we assume that touristic cuisines are an attempt to meet either of those goals, they cannot help but fall short. Restaurant food and domestic food are never really the same, and so food served to tourists in restaurants can never be exactly what locals are eating at home. Recreating cuisines in new social contexts must inevitably lead to accommodations. Instead, it is more productive to think about how touristic cuisine responds to new cultural logics in the "borderzone" of tourism (Bruner 2005), logics that draw on encounters between multiple groups of locals and tourists. Foods served to tourists reflect the cultural meanings produced by tourism itself as a cultural practice.

Cuisine is not a single, clearly demarcated thing. Few tourists are likely to believe—for most are not naïve—that what they consume in an expensive French restaurant in Paris is really what French people eat at home, or what they eat at a tourist restaurant in Cancún is the same as what Yucatán agriculturalists are eating. And yet, tourists are often asked to believe that at least some of what they are served in restaurants is somehow the "same" as what local residents are eating at home. In other words, they are asked to believe certain dishes are "authentic." This equation is in no way natural; ongoing cultural work must constantly recreate this equivalence between domestic/local cuisines and restaurant cuisines. The authentic must be made.

One meal does not define a cuisine; cuisines are the culinary structures that make meals make sense. These are the rules by which ingredients, spices, cooking techniques, and presentations are brought together, the understandings about how diners should interact with food in specific settings. What we label as "French" or "Mexican" cuisines are collections of diverse ingredients, cooking techniques, labor practices, dining etiquettes, and cultural knowledge, many of which have additional associations with economic class, regional identity, and other social markers. How does a diverse set of cooking techniques, ingredients, and dishes—albeit perhaps with a "family resemblance"—become recognized as a national or regional cuisine that can be discussed as an entity?

Recipes and cookbooks are one of the places where culinary rules are written, codified and disseminated (Appadurai 1988, Pilcher 1996, Ferguson 2010). We can also include forms of "secondary orality" (Ong 1982) such as food photography, cooking TV shows, visual blogs, and restaurant menus that contribute to defining and diffusing cuisines to larger audiences. Regardless of whether cuisines are presented as written recipes, mouth-watering photographs, or dishes at the table, culinary structures exist only in their physical manifestations. Like oral traditions, these structures are never set in stone. There is always interplay between cultural structures and their manifestation through particular food events that changes both.

This process of creating cuisine can lead to what I will call "haute traditional" cuisines. These cuisines bridge between a geographic localism and a globalizing cosmopolitanism. Local foodways imply those that are grounded in specific spaces, employ locally-produced or "native" ingredients, claim to be descended from the ancestral past, are produced by locally trained people using traditional techniques, and associated with domestic spaces. As such, such local foodways are often associated with the labor of unpaid (or underpaid) anonymous women cooking at home or in liminally public spaces such as street markets (Mangan 2005, Seligmann 1989, Weismantel 2001). Cosmopolitan cuisines, in contrast, suggest foodways that are perceived to cross international boundaries. These are produced by professionally trained chefs, largely men, who are paid and respected as individuals for their skills. Such cosmopolitan foodways are often presented and consumed in elite public spaces such as upper-end restaurants and public banquets, rather than in private homes.

Of course, this division is idealized. Most domestic cuisines employ ingredients or techniques that are not local, depending on how local is defined. But "haute traditional" cuisines are distinct because they explicitly move between the two extremes of this idealized division—local/native/ancestral/feminine vs. cosmopolitan/transnational/innovative/masculine—in order to claim legitimacy both as heritage cuisines *and* as a global elite commodities. In many cases, by doing this they also both re-inscribe and bridge divides in class, for example, making cosmopolitan elite fare from dishes associated with poverty into meals that the poor could never afford (as in the case of llama meat, to be discussed below). Similarly, "haute traditional" cuisines may bridge divisions of ethnicity, such as fusion cuisine that incorporates elements of distinct geographical foodways into new forms. The formation of these cuisines is not a natural process; cultural work goes into naturalizing haute traditional cuisines as foodways that can be simultaneously local and cosmopolitan.

I argue that the inscription of particular cuisines on the UNESCO List of Intangible Heritage is part of the process that creates "haute traditional" cuisines by linking elements perceived as "local" with those seen as "cosmopolitan." The process of nominating and honoring these cuisines does not simply identify pre-existing foodways, but actively creates links across the divides of the local and cosmopolitan, forming cuisines that can be both "haute" and "traditional" at the same time. This process is transformative, following the theoretical argument of Trouillot (2000), linking into a single cuisine distinct foodways that in other contexts might be seen as divided by class, ethnicity, or context. Through rituals such as international recognition as intangible heritage, diverse national foodways are transformed into cuisine.

This transformation is never permanent, however. Internal differences between local and cosmopolitan foodways are never truly erased, and therefore the transformative rituals that create haute cuisines must be continually renewed through tourism and heritage practices. While this process is by its very nature never complete, it is essential to creating and recreating "haute traditional" cuisine.

In this piece I will consider three examples—two of cuisines designated as UNESCO Intangible Heritage in 2010 (French and Mexican), and one of the incorporation of a local ingredient into international touristic fare (Thai llama curry in Bolivia). The first two will illuminate how the heritage and tourism "-scapes" work to create haute traditional cuisines by linking domestic cuisines to food served in restaurants. The process of nominating and then celebrating these cuisines as intangible heritage carefully constructs and naturalizes the connections between the local and cosmopolitan. In the case of Thai llama curry, we will see how an ingredient constructed as "good to think" in the context of Bolivian touristic cuisine became incorporated into a cosmopolitan cuisine catering to tourists. In an opposite move from the intangible heritage cuisines, in this case a cosmopolitan cuisine was localized itself by incorporating an ingredient that ia not only produced nationally, but was marked as particularly local in the context of Bolivian touristic cuisine.

Haute Traditional Cuisines on UNESCO's List of Intangible Heritage

The List of Intangible Heritage was intended to correct a perceived bias towards European and monumental heritage in the World Heritage List (Aikawa-Faure 2009: 13–44, Smith and Akagawa 2009: 1–10). Hafstein (2009), in his description of the inner workings of creating the List, gives an ethnographic description of how delegates argued about whether the Intangible Heritage List should be more or less inclusive, but notes that lists by their nature itemize, order, and exclude; they are "a tool for channeling attention and resources to certain cultural practices and not to others" (2009: 108).

Official criteria for inclusion on the List of Intangible Heritage focus on the locality of the heritage in question (UNESCO 2003: Article 2). Cuisines must be demonstrably local, grounded in historic agricultural and culinary practices, and practiced in household contexts. Cuisines included on the List were presented in their formal nominations as local and "authentic," grounded in practices that are domestic, semi-private, and largely feminine. The nominations downplayed the global importance of these cuisines, such as their global trajectories via tourists, professional chefs, and restaurants. These cosmopolitan elements came to the fore after the cuisines were named to the List, in celebrations of the UNESCO recognition of these newly-minted heritage cuisines and their subsequent marketing as national touristic attractions. Thus the UNESCO List of Intangible Heritage became part of the process of linking the local cuisines that are described in formal nominations to the List with the restaurant and touristic cuisines that create their public presentations. This is an active process that forms the basis for "haute traditional" cuisines, which must be simultaneously grounded in local foodways and appeal to global communities of elite diners.

It is interesting that the formal nominations of cuisines to the List minimize their cosmopolitan aspects, even though all those nominated are foodways with global reach. The three cuisines first recognized by UNESCO in 2010—French, Mexican, and Mediterranean—represent nations that are major tourist attractions, and thus have foodways known to a wide international audience. Five of these seven nations were in the top 10 in terms of absolute numbers of global tourists: France (1), Spain (2), Italy (5), and Mexico (7). Greece was listed at 17, and Morocco was listed as one of the "World's Top Emerging Tourism Destinations" for 1995–2004 due to the rate of increase of tourist arrivals (WTO 2004a, 2004b). It is noteworthy that the Peruvian government's campaign to have UNESCO recognize Peruvian cuisine (expected in November 2012) came with the tagline, *"Cocina peruana para el mundo"*—Peruvian Cuisine for the World. Acquiring UNESCO recognition thus explicitly forms part of a project to promote tourism through food (Cocina Peruana para el Mundo n.d., Wilson 2011).

Cuisine—like other forms of heritage—becomes meaningful in a global "heritage-scape" (Di Giovine 2009) in part through interactions with tourists. National cuisines cannot be merely of local importance if they are to appeal to foreign tourists and global restaurants. They must be diverse, internationally influenced, and appeal to a cosmopolitan audience. Thus "traditional" cuisines become recognized as "heritage" cuisines by being tied into the class-marked cuisine of the global transnational elite (for example, see Wilk 2006). This feeds back into touristic marketing; it is not a coincidence that national tourism boards are among the first to proclaim UNECSO honors for local cuisines. All of this contributes to naturalizing haute traditional cuisines as containing both local and cosmopolitan elements.

French Cuisine: Eating the Haute at Home

In 2010, the "French gastronomic meal" was added to UNESCO's List of Intangible Heritage. On the surface, this appears unsurprising. French cuisine dominates the fine dining of the industrialized world and the institutions that train chefs to participate in those networks (Ferguson 2010). Anthropologist of French cuisine Trubek writes, "the French invented the cuisine of culinary professionals" (2000: 3). French "haute cuisine" emerged beginning in the medieval era in the homes of the French nobility, where elite banquets were essential to creating and maintaining status. Elaborate presentations, exotic spices, and rare ingredients were important elements of these feasts. These elements continued to be central as French haute cuisine entered the public spaces of restaurants and hotels in the early nineteenth century (Trubek 2000: 3–8). French food is globally famous because it has influenced so many of the elite culinary practices of the restaurant world—in other words, the realm of cosmopolitan cuisine.

The UNESCO nomination for the French meal is interesting precisely because it says almost nothing about the international renown of French *haute cuisine* (UNESCO 2010a). It makes no mention of specific dishes, culinary techniques, or specialized culinary knowledge. Instead, the "gastronomic meal" described in the UNESCO nomination focuses on the social aspects of commensality: "The gastronomic meal is a social practice bringing together a group to mark, in a festive way, important moments in the lives of individuals and groups" (UNESCO 2010a: 2). The details of the food itself are specifically excluded: "This social practice is associated with a shared vision of eating well, rather than with specific dishes" (UNESCO 2010a: 3–4). There is little mention in of any specific ingredients or cooking techniques, other than general comments about a preference for local foods and a diversity of recipes. The nomination instead focuses on elements such as the presence of wine, the order of courses, the presentation of the table, and "codified gestures" for discussing and appreciating the food presented (UNESCO 2010a: 5). Meals are thus presented as rituals, involving celebrations of life events, songs, and commensality, largely independent from the material aspects of cooking itself.

The visual material on this UNESCO nomination presented on the UNESCO website (UNESCO 2010a) also focuses on the social aspects of French meals. Most of the images in the photo slideshow are of diners seated at tables of varying formality. Even where food is present in these photos, it is not the center of attention. A video of over nine minutes begins with images of the iconic French landscape and people laughing, dining, and eating in domestic and restaurant settings. The next section shows people buying fresh local ingredients, cheeses, and wines at market stalls and small stores, and commenting on the importance of *terroir* (although few images of agriculture are shown). The last half of the film shows a French family preparing a meal at home, emphasizing how parents teach children cooking skills. Although the meal is elaborate, the courses beautifully presented, and the table laid formally, this is not presented as a meal created by professionals for the commercial sphere, but rather one prepared and consumed in the domestic sphere.

Food as an important part of social interactions, a marker of important occasions, and a skill passed from parents to children all describe human universals, rather than elements unique to French cuisine. Culinary professionals, however, do learn specific culinary techniques that distinguish French cuisine, such as "stocks, sauces, knife skills, cooking methods, and pastry" (Trubek 2000: 13–30), in addition to rules governing the presentation of courses. But the skills of professionally-trained chefs are not mentioned in the UNESCO nomination as part of what makes French cuisine unique. Instead, the nomination focuses on the experience of the French as domestic preparers and consumers of food, marking "the French" as a community whose identity is created through French cuisine. This is true, but not exclusively so; social interactions and transmission of culinary knowledge also take place (albeit in distinct ways) in the transnational community of professional chefs trained in French culinary techniques. While the identity of professional chefs is also created through the acquisition of culinary skills, this group is not defined by the borders of the French nation-state. In fact, many chefs trained in the French tradition work outside France, preparing food that might not be labeled as French, as we shall see below in reference to Mexican cuisine.

Although the world of professional chefs is not mentioned in the UNESCO documents, it is difficult to imagine that French cuisine was honored only because it was important to the French. After all, culinary practices are everywhere important to those who employ them, yet most are not recognized as Intangible Heritage. French cuisine is widely equated with elite culinary culture, and that no doubt had much to do with the "French gastronomic meal" winning UNESCO recognition. French President Sarkozy, who was involved in the effort to have French cuisine included in the List, was widely quoted in 2008 as declaring "We have the best gastronomy in the world" (AFP 2008, Sciolino 2008) and therefore the French meal was an obvious choice. Yet when UNESCO made the decision to include the French Gastronomic Meal in the List, Francis Chevrier, director of the Institut Européen d'Histoire et des Cultures de l'Alimentation (European Institute of Food History and Culture), commented:

> It's very important that people realise, in villages in Africa and everywhere, that when you have knowledge of food it is a treasure for your community, and something worth cherishing … This is excellent news for French culture, for French heritage, to invite our cuisine, our gastronomic heritage to sit at the high table of culture at UNESCO. (Bachorz 2010, Samuel 2010)

French cuisine is simultaneously equivalent to a locally cherished village cuisine, equal in importance to any other, *and* sitting at "the high table of culture." The slippage between the local and the cosmopolitan is evident.

While the nomination process emphasized the former aspect, celebrations of the 2010 honor focused on the latter, linking the domestic French food in the UNESCO nomination to the cosmopolitan world of elite French chefs and restaurants. In April 2011 in Versailles—itself a World Heritage site—professional chefs gathered to celebrate the 2010 honor by feting 650 black-tie guests in the finest in French food. While some felt they fell short of high expectations (Sciolino 2011), this was hardly the domestic scene painted in the Nomination to UENSCO. It is precisely this tension between local and cosmopolitan that makes haute traditional cuisine "good to think" as a form of intangible heritage.

Michoacán/Mexican Cuisine: Making the Indigenous National

Michoacán/Mexican cuisine was also inscribed in the UNESCO List of Intangible Heritage in 2010. Michoacán, a state on the western Pacific coast of Mexico, is home to the Purépecha indigenous group, but this culture is only mentioned in passing in the nomination. Instead, Michoacán cuisine is referred to as "traditional," prepared by "traditional cooks" or occasionally "native peoples." Specific cultural populations living in Michoacán are not discussed as culinarily unique within the region. In addition there is slippage between national and regional designations in the nomination, itself titled "Traditional Mexican cuisine—ancestral, ongoing community culture, the Michoacán paradigm" (UNESCO 2010b). Although foodways in Mexico are culturally and culinarily diverse, the nomination moves between describing its object as "Michoacán cuisine" and "traditional Mexican cuisine," with the former being treated as representative of the latter.

This nomination differs from the French nomination in it has a lot more to say about actual food. Images of mouth-watering foods, both cooked and raw, dominate the visual materials in the UNESCO nomination. Michoacán/Mexican cuisine is presented through the trifecta of maize, beans, and chiles. Specifics such as cooking maize with limewater (*nixtamalization*), farming techniques such as forest clearing and intercropping (*milpas*), floating intensive fields (*chinampas*), and rituals where food is particularly important (Day of the Dead) are all specifically mentioned. In the nomination to UNESCO, essential elements of Michoacán/Mexican cuisine are those native to pre-Columbian Mesoamerica, such as cocoa, avocados, tomatoes, sweet potatoes, squashes, and especially maize. In the video accompanying the nomination, maize appears in the field, in contemporary and pre-Columbian artwork, dried, and as grain, tortillas, tamales, and soups. All cuisines are defined more by some elements than others; some ingredients, techniques, or presentations are essential, while others are transitory, ornamental, or irrelevant. In this case, maize is clearly "good to think" about what makes Michoacán/Mexican cuisine unique, a fact that has a long historical precedent (Pilcher 1996).

Throughout the nomination, Michoacán cuisine is linked to the indigenous past and indigenous women. Mexican cuisine in shown as both ethnically marked and gendered. Indigenous women dressed in ethnically-marked clothing are shown throughout the nomination document, and in the photographs that accompany it, at the UNESCO website. While many of these photos focus on the food itself, indigenous women dominate as the faces of traditional Michoacán cooks. These women appear in domestic and market settings. In these visual representations of Michoacán cuisine men are virtually absent, appearing briefly as either farmers or consumers of food. This gender disparity is also reflected in which community members signed in support of the nomination. Of those signing as "*herederas*"—specifically *female* inheritors of the tradition—45 were women, 10 were men (the gender of five could not be determined). Of those signing as representatives of organizations, NGOs, and other institutions, 32 were women, 11 were men. This document presents Michoacán/Mexican cuisine as pre-Columbian, indigenous, and feminine.

While tourists do not appear in any of the UNESCO documents, their presence is essential: they are one of these materials' intended audiences. The photos on the UNESCO website are all copyrighted by A. Rios/Secretaría de Turismo del Estado de Michoacán, the Tourism Office of the State of Michoacán. The Secretary of that body, Genovevo Figueroa Zamudio, is one of the signatories on the application. The UNESCO designation of Michoacán/Mexican cuisine was quickly heralded by the Mexico Tourism Board in order to promote Mexican cuisine through international restaurants, especially in those nations that send large numbers of tourists to Mexico. The Board organized a tour of Mexican chefs to U.S. and Canadian cities to showcase Michoacán cuisine (PRNewswire 2010).

The chefs chosen to represent this cuisine abroad were not indigenous women, but male Mexican chefs trained in the French culinary tradition—in other words, members of the professional culinary elite. I will give two examples here. Richard Sandoval, a prominent chef from Mexico City, was one of those chosen to showcase Michoacán cuisine to international audiences. He hosted nine dinners in December 2010 to celebrate UNESCO's decision to honor Mexican cuisine, all in the United States. Sandoval's resumé is impressive. Around the time of this 2010 tour, he had 22 restaurants, 10 of which he had opened within the previous 15 months (Jennings 2010). He and his restaurants have won numerous awards, all of which are listed on his website (www.richardsandoval.com), including being named the Torque d'Oro Mexican Chef of the Year in 1992. Sandoval is an innovator of Modern Mexican Cuisine, about which he has published a cookbook (Sandoval et al. 2002). A 1991 graduate of the Culinary Institute of America, Chef Sandoval is versed in French and other culinary forms. Before opening modern Mexican restaurants, he opened two restaurants in New York City offering modern French cuisine (Culinary Institute of America n.d.). Several of his restaurants feature Latin-Asian fusion cuisine, and he emphasizes combining Mexican ingredients with European cooking techniques. Sandoval teamed with American Airlines in 2011 to develop for-purchase meals "using authentic Latin ingredients" for premium class passengers on international flights (American Airlines n.d., 2011). Ironically, that same year, he told the Zagat blog that what he wanted for the holidays was an airplane—a playful reference to his travels between his restaurants in the United States, Mexico City, and the Middle East (Covington 2011). In short, Sandoval is well-established in creating an innovative Mexican cuisine that appeals to an elite transnational audience.

Another Mexican chef involved in the UNESCO celebrations was Carlos Gaytán, who opened *Mexique*, an award-winning Chicago restaurant, in 2008. Mexique advertises itself as offering a "historical blending of French and Mexican cuisine" through the expertise of Gaytán, who "combines his Mexican heritage with French training for a sophisticated result" (Mexique Restaurant n.d.). Gaytán's 2010 meal sponsored by the Mexican Tourism Board began with *ceviche*, a dish of raw fish "cooked" in lime juice typically associated with Peruvian coastal cuisine, garnished with chipotle aioli, avocado mousse, and mango habanera galette. This was followed by a dish entitled *"Mole y Maíz,"* made with Jamaica Braised Pork Belly, and *"Carne Asada"* served with fingerling potato salad and goat cheese. These dishes were followed with tequila and chocolate (Just Get Floury 2010). Gaytán's meal combines pre-Columbian Mexican ingredients (avocado, maize, chocolate), ingredients now common in Mexico foodways (beef, mango, pork) and culinary techniques and ingredients that reference other places (galette, ceviche).

This combination of ingredients, cooking techniques, and presentations is not unusual in high-end restaurants, nor is this unusual in elite Mexican cooking, which has long incorporated international ingredients and techniques (Pilcher 1996). While the UNESCO nomination for Michoacan cuisine focused on the localness, indigeneity, and domesticity (via female cooks) of Michoacan's cuisine, its promotion to an international audience focused on a cuisine connected to a global networks of ingredients, cooking techniques, professional chefs, and international diners.

Obviously Mexican cuisine has both these elements—roots in ancient pre-Columbian agricultural and cooking techniques, and global influences and audiences. My purpose is not to argue that one kind of Mexican cuisine is more authentic—whatever that might be interpreted to mean—than the other. These are both Mexican cuisines, prepared in particular social contexts for particular audiences, albeit quite different in terms of their processes of production, contexts of consumption, and intended audiences. What is interesting is how naming "Michoacán/Mexican" cuisine as UNESCO Intangible Heritage involves a clear effort to actively link these distinct foodways together as part of the same cuisine. The fact that the designation of "Mexican cuisine" might seem a self-evident phrase to describe both of these foodways shows the success of "haute traditional" in linking the local and the cosmopolitan.

From Local to Cosmopolitan and Back Again: Llama Curry

It is not only internationally-popular national cuisines honored as Intangible Heritage that actively emphasize their local connections. I will now consider the example of an international restaurant that highlights how it incorporates local ingredients into its foreign dishes. While this does not involve the Intangible Heritage List directly, I argue that the same tensions are present between presenting foods as both local and cosmopolitan. This suggests that the criteria that govern how cuisines become labeled as Intangible Heritage bear similarity to those that make for felicitous touristic cuisine. Both must link the local and the cosmopolitan.

This set of ethnographic observations takes us to Sagarnaga Street, the touristic district of La Paz, Bolivia. Bolivian cuisine is not one with a wide international following, and touristic discourse about the region rarely focuses on its food as a major attraction. This is unfortunate neglect, in my opinion, for the cuisine that includes such treats as *chairo* (soup made from ground freeze-dried potatoes), *salteñas* (large pastries stuffed with juicy beef stew, a delicacy I have never learned to eat neatly), and *sajta de pollo* (a spicy chicken dish).

The Bolivian tourism industry focuses on indigenous cultures in the highlands (Aymara and Quechua) and ecotourism in the Amazonian lowlands. Government and popular touristic materials declare the nation to be the "*Capital de Folklore de Suramérica*" (Folklore capital of South America) or a place where "*Lo auténtico aún existe*" (The authentic still exists). But little of this interest is focused on food. While UNESCO's List of Intangible Heritage includes Aymara culture, that nomination says nothing about food, instead concentrating on music, language, and to a much lesser extent agricultural practice (UNESCO 2009). The potato, native to the Andes, is now central to many of the world's cuisines, but despite this—or perhaps *because* of this (Sammells 2010)—few think of Bolivian highland cuisine as an attraction in and of itself.

Even if tourists don't come to Bolivia for the food, they must still eat. As in many other touristic districts, Sargarnaga Street in La Paz caters to tourists with a collection of restaurants, hotels, hostels, and nearby cultural attractions such as the San Francisco Cathedral and the "Witch's Market." Like many other urban touristic spaces, the restaurant offerings of the Sargarnaga district are international, including "Bolivian" fare, American food, European dishes, vegetarian meals, Israeli cuisine, and numerous bars.

The "Bolivian" food offered in Sagarnaga is not representative of what most Bolivians eat at home, however. Bolivia's touristic cuisine has its own culinary grammar that merges elements of local foodways, tourist preferences, and local understandings of tourists' desires and social roles to create a new type of cuisine with a specific set of meanings. In this context, certain local ingredients are seen as more Bolivian than others. Llama meat has become an increasingly visible part of the cuisine offered to foreign visitors. The animal appears on postcards and national insignias as a symbol of the Andes. Llama meat is closely associated with indigenous peoples and poverty, but since 1997 it has become common on the menus of touristic restaurants in La Paz, first appearing in a touristic *peña*, a type of touristic restaurant that offers Bolivian cuisine and a show of traditional Bolivian dances, largely for foreign visitors. In this context, llama meat has become a symbol of "authentic" Bolivian fare, and is advertised prominently on signs outside many establishments (Figure 9.1). Llama meat is rarely seen in non-touristic restaurants in La Paz; indeed, in the early 1990s no La Paz restaurants would have served this kind of meat, and even now llama meat is rare in La Paz establishments outside the tourist district (Sammells 1998, 2010).

Llama meat is consumed by some rural indigenous highland Bolivians, but not all. In Tiwanaku, a village two hours outside of La Paz where I conducted my dissertation research, llama was never served in households. This region produces dairy, and farmers also keep sheep, pigs, chickens, and guinea pigs in addition to practicing agriculture. Beef and mutton were included in small quantities in soups, and pork was served for celebratory events. The one llama herd in the village was kept by the local university as part of their agricultural program. Those animals, when slaughtered, were processed and sold in markets in La Paz rather than consumed locally. Nevertheless, Tiwanaku's tourist restaurants proudly advertised and served llama meat, just like their counterparts on Sagarnaga Street. They purchased their llama meat in the city and transported it back to the village for sale. While llama meat evoked rural indigenous lifestyles in the minds of tourists, it tied Tiwanakeño touristic establishments into urban networks of food provisioning.

In conversations with Tiwanakeños about what foods they feel are truly *típico,* and in observing the offerings of local food fairs and household cuisine, other kinds of foods besides llama are presented as most symbolic of Andean cuisine. These include grains such as *quinua*, which is commonly served in touristic spaces in soups and has recently become a major export for Bolivia. Tiwanakeños also think about local cuisine in terms of dishes that I have never seen offered at any touristic restaurant, such as *pit'u* (powdered quinua, cañahua, maize, or fava beans

Figure 9.1   **Bolivian restaurant in the tourist district of La Paz advertises
llama meat steaks**

Figure 9.2 Thai restaurant in the tourist district of La Paz, Bolivia offers llama meat curry

stirred with hot tea into a thick porridge) and *wila parka* (a dish made with sheep's blood). For Bolivian touristic cuisine, llama has become the symbol of authenticity, the ingredient that marks meals as "truly" Bolivian, in part through the validation of networks of tourism. But this is not necessarily agreed upon by all Bolivians as being what constitutes "authentic" Bolivian food, nor is it a reflection of what many Bolivians regularly eat.

Despite these multiple understandings of Bolivian cuisine, by 2010 llama meat was entrenched as a ubiquitous part of highland Bolivian touristic fare. In contrast, Thai ethnic restaurants had been established in La Paz only within the previous two years; this cuisine was seen as relatively new and exotic in the region. One of the Thai restaurants located on Sagarnaga Street, interestingly, prominently advertised its llama meat curries. Llama could now be consumed by tourists as part of either Bolivian touristic cuisine, or as international fare (Figure 9.2). This is not merely a matter of using local ingredients; llama is less available than beef or mutton in La Paz markets, and since its incorporation into touristic cuisine it is no cheaper. Including llama on the menu, and prominently displaying this on the street for potential tourist customers, send the clear message that despite the "international" status of the cuisine offered, this restaurant still incorporates iconic Andean elements. For a Thai restaurant in Bolivia that wants to offer both a cosmopolitan international fare and authentically local dishes, llama curry provides a creative, syncretic answer.

Conclusion

French and Mexican cuisines have their roots in local foodways but also transcend them, both having been international for centuries. The UNESCO List of Intangible Heritage does not change that history, but it does transform and inscribe how these cuisines are conceptualized, commoditized, and marketed. UNESCO's nomination and inscription process acts as a ritual, linking together foodways into a single cuisine across divides of ethnicity, class, gender, and public/private eating. Nominations to the UNESCO List of Intangible Heritage present cuisines as rooted in local communities, but their importance as "intangible heritage" is also intimately connected to their appeal to global, cosmopolitan audiences. Through such processes, elite international restaurants claim to offer cuisine that represents an "authentic" version of national cuisine, connecting elite global restaurant fare to what people cook at home within the geographic boundaries of the nation.

As researchers, we cannot assume that what is local fare is a given set of foodway practices that is then simply copied or co-opted by restaurants. When defining cuisines as either intangible heritage or as touristic fare, locality cannot be assumed to be a simply fact. Defining the "localness" of cuisines in both heritage and touristic contexts requires evaluating those cuisines through the lens of cosmopolitanism; localness must be created, demonstrated, and proven. The inclusion of cuisines in the UNESCO List of Intangible Heritage is not just an acknowledgement that food is important to human experience; that is true of all food. These cuisines have been nominated and accepted to the list in part because they are already recognized internationally. Their importance already extends beyond their own national boundaries. They are already "haute;" it is the "traditional" that needs to be demonstrated, documented, and highlighted in order for such cuisines to be accepted as intangible heritage. Thus the UNESCO nominations focus on these local aspects, grounding these national cuisines in household practice.

This active process of linking local to cosmopolitan is also evident for many touristic cuisines. Restaurants that serve tourists are by nature cosmopolitan—they are public spaces that serve people traveling through international networks. Thus, it is locality that such establishments need to prove. How that locality is demonstrated to a European tourist in Bolivia is not necessarily the same as what would mark such locality to an indigenous Aymara person eating at home. The very things that prove localness are inherently wrapped up with how specific audiences understand and conceptualize the divides between local and cosmopolitan.

Cuisines can and do, of course, have both local meanings and international followings; they can exist in both domestic kitchens and cosmopolitan restaurants. What is important to remember is that tourism and designations of intangible heritage create spaces for new kinds of social encounters (Liechy 2001). These new forms of sociality change how participants think about what constitutes the local, and what the importance of being local—and being recognized as local—is.

References

AFP 2008. Sarkozy Wants French Gastronomy Declared a World Treasure. *AFP* [Online, 23 February] Available at: http://afp.google.com/article/ALeqM5gA9J8yihnxy2w79i1gmVKUNBsTJg (Accessed 6 February 2012).

Aikawa-Faure, N. 2009. From the Proclamation of Masterpieces to the *Convention for the Safeguarding of Intangible Cultural Heritage*, in *Intangible Heritage*, edited by L. Smith and N. Akagawa. London and New York: Routledge, 13–44.

American Airlines n.d. New Chefs. New Dining Options. [Online: American Airlines]. Available at: http://www.aa.com/i18n/utility/newDining.jsp (Accessed 27 January 2012).

American Airlines 2011. American Airlines Collaborates With Chef Richard Sandoval and Chef Marcus Samuelsson to Enhance Inflight Dining Program. [Online: American Airlines]. Available at: http://aa.mediaroom.com/index.php?s=43&item=3343 (Accessed 27 January 2012).

Appadurai, A. 1988. How to make a National Cuisine: Cookbooks in Contemporary India. *Comparative Studies in Society and History* 30(1), 3–24.

Bachorz, B. 2010. French cuisine named 'intangible' world heritage. *AFP*. 16 November, 2010.

Bruner, E.M. 2005. *Culture on Tour: Ethnographies of Travel*. Chicago and London: The University of Chicago Press.

Cocina Peruana para el Mundo n.d. Cocina Peruana para el Mundo. [Online] Available at: http://www.cocinaperuanaparaelmundo.pe (Accessed 6 February 2012).

Counhian, C. 1988. Female Identity, Food, and Power in Contemporary Florence. *Anthropological Quarterly* 61(2), 51–62.

Counhian, C. and Esterik, P.V. 1998. *Food and Culture—A Reader*. London: Routledge, 2nd ed.

Covington, L. 2011. Chefs' Wish List: What Do the Pros Want for the Holidays? [Online: Zagat Blog]. Available at: http://www.zagat.com/buzz/chefs-wish-list-what-do-the-pros-want-for-the-holidays (Accessed 27 January 2012).

Culinary Institute of America n.d. Richard Sandoval '91, Owner of Modern Mexican Restaurants. [Online: Culinary Institute of America]. Available at: http://www.foodislife.org/cia-alumni-profiles/richard-sandoval-91-owner-of-modern-mexican-restaurants (Accessed 27 January 2012).

Di Giovine, M.A. 2009. *The Heritage-Scape: UNESCO, World Heritage, and Tourism*. Lanham, Boulder, New York, Toronto, Plymouth UK: Lexington Books.

Dixon, J. and Jamieson, C. 2005. The Cross-Pacific Chicken: Tourism, Migration and Chicken Consumption in the Cook Islands, in *Cross-Continental Food Chains*, edited by N. Fold and B. Pritchard. London and New York: Routledge, 81–93.

Dunn, E.C. 2004. *Privatizing Poland: Baby Food, Big Business, and the Remaking of Labor*. Ithaca and London: Cornell University Press.

Ferguson, P.P. 2010. Culinary Nationalism. *Gastronomica* 10(1), 102–9.

Gill, L. 1994. *Precarious Dependencies: Gender, Class, and Domestic Service in Bolivia*. New York: Columbia University Press.

Ginny. 2010. Celebrating Mexico @ Mexique! [Online: Just Get Floury]. Available at: http://www.justgetfloury.com/?m=201012 (Accessed 27 January 2012).

Hafstein, V.T. 2009. Intangible Heritage as a List: From Masterpieces to Representation, in *Intangible Heritage*, edited by L. Smith and N. Akagawa. London and New York: Routledge, 93–111.

Jennings, L. 2010. Richard Sandoval tweaks concepts as he expands. *Nation's Restaurant News* [Online, 10 November]. Available at: http://nrn.com/article/richard-sandoval-tweaks-concepts-he-expands (Accessed 2 February 2012).

Liechty, M. 2001. Carnal Economies: Commodification of Food and Sex in Kathmandu. *Cultural Anthropology* 20(1), 1–38.

Mangan, J.E. 2005. *Trading Roles: Gender, Ethnicity, and the Urban Economy in Colonial Potosí*. Durham and London: Duke University Press.

Mexique Restaurant Webpage n.d. [Online: Mexique Restaurant]. Available at: http://www.mexiquechicago.com (Accessed 27 January 2012).

Ong, W.J. 1982. *Orality and Literacy: The Technologizing of the Word*. London and New York: Routledge.

Pilcher, J.M. 1996. Tamales Or Timbales: Cuisine and the Formation of Mexican National Identity, 1821–1911. *The Americas* 53(2), 193–216.

PRNewswire 2010. Mexico Tourism Board Celebrates UNESCO Honor for Traditional Mexican Cuisine. *PRNewswire* [Online, 22 November]. Available at: http://www.prnewswire.com/news-releases/mexico-tourism-board-celebrates-unesco-honor-for-traditional-mexican-cuisine-109871009.html (Accessed 25 January 2012).

Sammells, C.A. 1998. Folklore, Food, and Seeking National Identity: Urban Legends of Llama Meat in La Paz, Bolivia. *Contemporary Legend* 1: 21–54.

Sammells, C.A. 2010. Ode to a Chuño: Learning to Love Freeze-Dried Potatoes in Highland Bolivia, in *Adventures in Eating: Anthropological Experiences of Dining from Around the World* edited by H.R. Haines and C.A. Sammells. Boulder: University Press of Colorado, 101–26.

Samuel, H. 2010. UNESCO declares French cuisine 'world intangible heritage.' *The Telegraph* [Online, 16 November]. Available at: http://www.telegraph.co.uk/news/worldnews/europe/france/8138348/UNESCO-declares-French-cuisine-world-intangible-heritage.html (Accessed 6 February 2012).

Sandoval, R., Ricketts, D. and Urquiza, I. 2002. *Modern Mexican Flavors*. New York: Stewart, Tabori and Chang.

Sciolino, E. 2011. Sixty Chefs in the Palace, and Still 'Just Average.' *New York Times Diner's Journal* [Online, 8 April]. Available at: http://dinersjournal.blogs.nytimes.com/2011/04/08/sixty-chefs-in-the-palace-and-still-just-average (Accessed 8 August 2012).

Sciolino, E. 2008. Time to Save the Croissants. *New York Times* [Online, 24 September]. Available at: http://query.nytimes.com/gst/fullpage.html?res=9C03EEDA1E3BF937A1575AC0A96E9C8B63 (Accessed 8 August 2012).

Seligmann, L.J. 1989. To be in between: The *Cholas* as Market Women. *Society for Comparative Study of Society and History,* 694–721.

Skounti, A. 2009. The Authentic Illusion: Humanity's Intangible Cultural Heritage, the Moroccan Experience, in *Intangible Heritage*, edited by L. Smith and N. Akagawa. London and New York: Routledge, 74–92.

Smith, L. and Akagawa, N. 2009. Introduction, in *Intangible Heritage*, edited by L. Smith and N. Akagawa. London and New York: Routledge, 1–10.

Trouillot, M-R. 2000. Abortive Rituals: Historical Apologies in the Global Era. *Interventions* 2(2), 171–86.

Trubek, A.B. 2000. *Haute Cuisine: How the French Invented the Culinary Profession.* Philadelphia: University of Pennsylvania Press.

UNESCO 2003. Text of the Convention for the Safeguarding of Intangible Cultural Heritage. [Online: United Nations Educational, Scientific and Cultural Organization]. Available at: http://www.unesco.org/culture/ich/index.php?lg=en&pg=00022 (Accessed 2 February 2012).

UNESCO 2010a. The Gastronomic Meal of the French: Nomination File no. 00437 for Inscription on the Representative List of the Intangible Cultural Heritage in 2010 [English] [Online: United Nations Educational, Scientific and Cultural Organization]. Available at: http://www.unesco.org/culture/ich/index.php?lg=en&pg=00011&RL=00437 (Accessed 6 August 2012).

UNESCO 2010b. Traditional Mexican cuisine—ancestral, ongoing community culture, the Michoacán paradigm: Nomination File no. 00400 for Inscription on the Representative List of the Intangible Cultural Heritage in 2010 [English] [Online: United Nations Educational, Scientific and Cultural Organization]. Available at: http://www.unesco.org/culture/ich/index.php?lg=en&pg=00011&RL=00400 (Accessed 6 August 2012).

Weismantel, M.J. 1988. *Food, Gender, and Poverty in the Ecuadorian Andes.* Philadelphia: University of Pennsylvania Press.

Weismantel, M.J. 2001. *Cholas and Pishtacos: Stories of Race and Sex in the Andes.* Chicago: University of Chicago Press.

Wilk, R.R. 1999. "Real Belizean Food": Building Local Identity in the Transnational Caribbean. *American Anthropologist* 101(2), 244–55.

Wilk, R.R. 2006. *Home Cooking in the Global Village: Caribbean Food from Buccaneers to Ecotourists.* Oxford and New York: Berg.

Wilson, R. 2011. Cocina Peruana Para El Mundo: Gastrodiplomacy, the Culinary Nation Brand, and the Context of National Cuisine in Peru. *Exchange: The Journal of Public Diplomacy*, 13–20.

WTO 2004a. World's Top Emerging Tourism Destinations for period 1995–2004. [Online: World Tourism Organization]. Available at: http://www.unwto.org/facts/eng/pdf/indicators/ITA_emerging04.pdf (Accessed 6 February 2012).

WTO 2004b World's Top Tourism Destinations (2005). [Online: World Tourism Organization]. Available at: http://www.unwto.org/facts/eng/pdf/indicators/ITA_top25.pdf (Accessed 6 February 2012).

Chapter 10

Reinventing Edible Identities: Catalan Cuisine and Barcelona's Market Halls

Josep-Maria Garcia-Fuentes, Manel Guàrdia Bassols
and José Luis Oyón Bañales

Our ongoing research on the history of market halls, sponsored by the Spanish government,[1] highlights how the recent renewal of these facilities in Barcelona coincides with the emergence of other processes involving the relationships between identity and food, such as the recovery of Catalan cuisine, the advocacy of the Mediterranean diet and the eating practices of new immigrant groups. Although these processes may appear to be totally independent, their development shows similarities that can hardly be attributed to chance. This chapter explores some of the most important aspects that have arisen in recent decades while looking into the changes in the context and meaning of the market halls and, more generally, of food. To this end, this study is based on the documentary sources of the city council's management bodies and of the market halls themselves, and on newspapers, articles and more popular materials, including television programs. This diversity of documentary sources allows different materials to be combined, showing how a new frame of reference that differs completely from the previous one has been created.

The study of food identities can be attributed to the growing interest in food and eating (Watson and Caldwell 2005) and to the broader field of consumer patterns as an instrument of social differentiation and identity building (López de Ayala 2004). Additionally, the fact that food identities are regarded as cultural heritage sets them within the sphere of globalization and emphasizes processes within the "local production structure" (Appadurai 1996: 178), the establishment of a "common local culture" (Featherstone 1991: 217) or, in stricter economic terms, the formation of a local "monopoly rent" (Harvey 2002).

Although the subject of identity has been a key political issue in the Spanish autonomous region of Catalonia for some time now, the unexpected situation of food identity has posed some interesting questions. The relationship between the recovery of traditional Catalan cuisine and the renovation of Barcelona's network of market halls is particularly significant. These are two processes that, while

---

1   MICINN, Plan Nacional I+D+i, Reference: HAR2009-09227.

apparently reflecting different independent logics in terms of both their objectives and their social agents, are doubtless reactions to the globalization process that had been associated with the fast food industry until the end of the 1970s (Watson and Caldwell 2005: 2). The relationship between market policies and the various initiatives involving food identities shows a by no means accidental convergence resulting from a common network of issues and meanings. It illustrates the fact that concepts such as identity or heritage are not something static and final, but rather open processes full of interactions that weave and consolidate networks of shared meaning, forming a symbolic capital characterized by a remarkable solidity and autonomy as well as by a notable plasticity.

The Erosion of Typical Cuisine

The remark attributed to the celebrated Catalan writer Josep Pla, "cooking is the landscape in a pan" (Santamaria 2005), clearly illustrates the notion that traditional cuisine is born from the logic of proximity. In Catalan towns and cities, the historical intermediary of this fusion of cuisine and landscape was the weekly market. There, the produce of the immediate environment converged, creating local dishes based on fresh market products. Each region identified itself with a certain landscape, and each landscape with a locality or cuisine. Landscape and cuisine consequently became two of the most widely shared expressions of local identity. In the last half of the nineteenth century, Barcelona emerged as the port of entry of more cosmopolitan influences derived from international cooking styles based mainly on French haute cuisine, which were to enrich and complement the local culinary traditions (Luján 1993). Accordingly, it is not surprising to find that, as far as the chroniclers of Catalan cuisine were concerned (Luján and Perucho 1970, Pla 1972, Luján 1977, Vázquez Montalbán 1977), this Golden Age gradually fell into decline in the twentieth century owing to a misconceived modernization and the slow but inevitable erosion of local cooking.

As what usually happened in Spain and other Western European countries, this evolution was the result of new methods of transport and increased urban growth that brought the logic of proximity to an end, as new forms of distribution displaced many products that had formed the basis of traditional cooking (Guàrdia and Oyón 2010: 62). In the process, local market halls gradually lost their important central role in the distribution system and, by the end of the 1950s, they were perceived as mere anachronistic vestiges of the past. This sped up the expansion of supermarkets, self-service stores and usually—following the American model—major discount chains that spread throughout Western Europe and led to the "pre-packaging of goods on a scale never before seen" (Schmiechen and Carls 1999: 209).

The Depression in the 1930s, the Spanish Civil War, and the poverty and rationing in the long post-Civil War period also produced an abrupt impoverishment of the Spanish diet. The uneven economic growth of the 1960s did not lead to any improvements but rather to the abandonment of many native crops, considered low-yielding, that were replaced by others that were quicker and less costly to grow. Moreover, new, cheaper products were promoted, such as sunflower oil instead of traditional olive oil. It also implied the progressive standardization and pre-packaging of products and a growing preference for faster and more convenient food preparations, reflected in the transformation of the traditional corner grocery shop into a self-service store. Overall, neither economic growth nor the new expansion of the tourism sector brought about any improvement in dietary standards in terms of diversity or better nutrition, and much less any recovery of traditional cooking. They did, however, generalize a handful of simple monotonous recipes that served to standardize contemporary cookery (Vázquez Montalbán 1977: 21).

Nevertheless, a marked difference may be observed in Spain with respect to other European countries. While the number of public markets has decreased throughout Europe because of the expansion of new commercial forms, in Spain the attempts to modernize the commercial distribution systems clashed with the country's socioeconomic reality. The commercial networks' fragility went hand in hand with poorly equipped households—the late adoption of electrical appliances (including the electric refrigerator), which was a sign of the small proportion of women in the workforce, and the likewise belated emergence of car ownership. As a result, it was necessary to rely on market halls as the only effective way of keeping prices in check, and the Barcelona City Council's active policies in relation to these facilities doubled their number between 1939 and 1980 (Guàrdia and Oyón 2010: 289). This unusual expansion and vitality of the market hall system reflected Spain's relatively underdeveloped economy as compared to other countries and was a fundamental starting point for this system's modern reinvention aimed to restore traditional shopping and dietary practices.

Catalan Cuisine and Barcelona's Market Halls

In the 1970s, at the peak of the anti-Franco resistance movement and the militant defense of the Catalan language—the backbone of Catalonia's claims to identity—the well-known mystery writer Antonio Vázquez Montalbán, in his *L'art del menjar a Catalunya* (*The Art of Eating in Catalonia*), published in 1977, established an ironic and provocative if not heretical culinary parallel to the mission of protecting the language. This parallel is clearly underscored in the subtitles of the book's various re-editions: *Crònica de la resistència dels senyals d'identitat gastronòmica catalana* (*Chronicle of the Resistance of the Distinguishing Traits of Catalan Cuisine*) published in 1984 or, in a more recent edition from 2004, *El llibre roig de la identitat gastronòmica catalana* (*The Red Book of Catalan Culinary Identity*).

Although turning cuisine into a component of national identity may have seemed a joke at that time, it marked a turning point. The 1980s saw an extraordinary renewal that can be indirectly measured by tracking the sudden increase in the number of journalistic references to the term "Catalan cuisine" found in *La Vanguardia*, a major Spanish newspaper published in Barcelona. Vázquez Montalbán himself, in the prologue to his second edition of the *L'art de menjar a Catalunya* published in 1984, acknowledges that, while in 1977 many of the recipes were mainly archaeological vestiges, the culinary panorama had radically changed seven years later, witnessing how much the book had contributed to the revival of Catalan cooking. This revival coincided with the resolute championing of market cuisine and quality fresh produce, which were also among the better-known characteristics of French nouvelle cuisine, a form of cooking that was at the height of its popularity at that time. In 1976, Paul Bocuse published *Market Cuisine*, which has been republished on several occasions and is now a classic, but many years earlier he had already become famous for his defense of quality fresh produce and for his and other great French chefs' ritual of deciding on each day's menu on their early-morning visits to the market. Accordingly, it is no surprise to find that the references in *La Vanguardia* to the term "market cuisine" coincided closely with those made to "Catalan cuisine."

In both cases the influence of the French experience should be considered. In the 1970s, France was the scene of the emergence of nouvelle cuisine, which championed market produce, and of the implementation of the Royer Act of 27 December 1973, designed to protect local commerce from major supermarket chains (Arribas and Van de Ven 2003). These two factors were reactions to the invasion of foreign influences that were believed to be destabilizing traditional forms of cuisine and the commercial fabric. Critical damage could be observed in the loss of food quality and the destruction of an essential economic structure, with the associated danger of the desertification of urban centers. Although there are no direct links, it was hardly by coincidence that the gestation of a policy to revitalize public markets occurred at the same time as a culinary renaissance.

Barcelona's policies with respect to market halls in the 1980s were quite unusual, both nationally and internationally, and they should be considered in the context of the interventionist policies of Spain's democratic city councils during those years, driven by their new legitimacy, the need to respond to the shortcomings inherited from the Franco regime and a particularly severe and prolonged economic crisis. In contrast to the deregulatory tendencies characterizing Thatcherism in the UK and Reaganomics in the U.S., Spain's democratic city councils embarked on a resolute policy of "urban reconstruction" beginning in the 1980s. It was in this context that Barcelona implemented in 1985 its Special Plan for Retail Establishments, designated by the Catalan acronym PECAB. The plan was designed to redefine the city's network of small local retail shops that characterized Barcelona's streetscape and structured its neighborhoods but had declined in the crisis of the 1970s. The intervention focused on a large network of 41 market halls, which were fully active at the time, where the city's families already spent a large portion of their food budget and which were evenly distributed over a relatively small area (PECAB 1990).

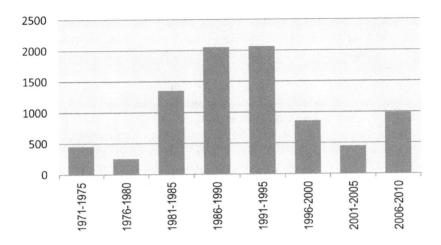

Figure 10.1   Number of references to "Catalan cuisine" in *La Vanguardia*

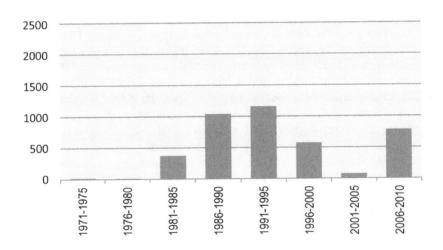

Figure 10.2   Number of references to "market cuisine" in *La Vanguardia*

This meant a radical rethinking of the market halls' main functions. At that time, market halls had the primary function of holding prices in check. The PECAB, however, saw them as efficient tools for fighting the effects of the rapid penetration of big shopping malls, which were causing many small retail shops across the city to close down and were changing the characteristic local urban space of neighborhoods and streetscapes. Any intervention on these key contrasting concepts had to endow market halls and their surroundings with efficiency and competitiveness in order to prevent the disappearance of local retailers and the dismembering of neighborhood life, as had happened in other more economically developed countries. In short, in the case of Barcelona, small retail shops and market halls became strategic elements in the preservation of the economic balance of the city's retail structure and, in turn, in the protection of the urban spaces associated with it. Consequently, retail shops and market halls became tools of economic and urban renewal.

The policy of reappraising and preserving the old wrought-iron market halls was consolidated in the 1970s with the debates on the demolition of Les Halles in Paris, the disputed conservation of Covent Garden market in London and, in the case of Barcelona, the spirited defense of El Born market. These debates established that the renewal of the market halls should also take clearly into consideration the heritage value of these buildings. It should be pointed out, however, that in the 1970s, this conservationist approach was not limited to architecture alone. The experience in Bologna (Cervellati and Scannavini 1976), which advocated the comprehensive conservation of the residential fabric in the neighborhoods of the historic center as well as the preservation of their social content with all their socioeconomic implications, was also a decisive influence and the approach taken by the PECAB paralleled that experience. Although it did not form part of the core of the argument at that time, there was still a latent implicit defense of native ways of life. The defense of traditional life, however, involved a steadfast modernization of its administration. The physical rehabilitation of buildings and the amelioration of distribution structures did not suffice to assure their competitiveness or to confront the new highly adaptable commercial forms that were emerging. The commonly used public administrative mechanisms were inefficient and excessively regulated, with scant funds available for investment in advertising and publicity. When it came to administering municipal market halls, the city council could hardly afford to limit itself to maintenance and conservation: it needed to apply modern commercial methods (*Memòria* 1991). Accordingly, in the first stage, together with a far-reaching architectural restructuring, it was decided to launch a major commercial promotion campaign and to spread the word about these market hall policies abroad through symposiums and conferences, which among other things allowed the incorporation of experiences gained from working with shopping centers.

With these precedents in mind, the Barcelona Municipal Institute of Market Halls, known by the Catalan acronym IMMB, was created in 1992 as an autonomous body attached to the City Council, in charge of the direct management and administration of the municipal markets. Over the past 20 years, this body has carried out a vigorous multifaceted policy of renewal and promotion. It has followed the guidelines laid down by the PECAB on the optimization of the retail mix, reducing the number of market stalls while increasing the levels of investment in remaining ones, and regulating the establishments in the surroundings of the market halls to ensure competitiveness. It also contributed to the physical renovation of the markets with the aim to adapt them to present-day functional requirements. The promotional campaign, however, has acquired an increased presence. The architectural interventions themselves have surpassed the operating requirements and have undertaken a decided redefinition of the old market halls' image through commissions to well-known architects, on occasion with spectacular results (as in the case of the Santa Caterina Market Hall, designed by the architect Enric Miralles).

Even more active and effective has been its capacity to multiply initiatives and revise the reasons for promoting municipal market halls. One of the Institute's main objectives has been the inclusion of new social and cultural uses inside the markets in close collaboration with the associative, neighborhood and educational fabrics of the surrounding areas. For example, cooking workshops and dietary education classes are now offered. This social and educational task has acquired a strong promotional component. The central motivations involve the close association of the municipal market halls with quality fresh produce; the spread

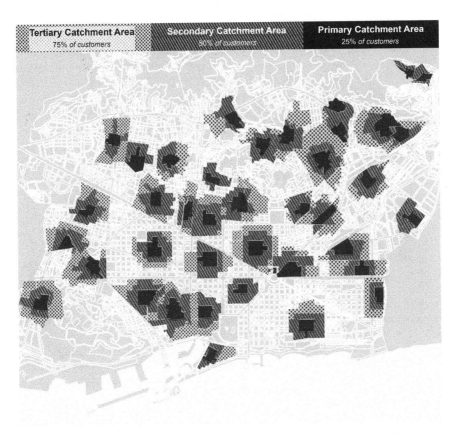

Figure 10.3   Clientele areas of Barcelona's zonal municipal market halls, 2011

of the principles of a healthy diet, and the promotion of quality cuisine based on a return to traditional food and the cooperation of great Catalan chefs. The city council's market policies have actively participated in an intense cultural and identity-based construction that has coalesced during these years around Catalan cuisine, market cuisine and what has come to be known as the "Mediterranean diet"—to such an extent that municipal market halls now present themselves as "cathedrals of the Mediterranean diet" (Duncan 2010).

The Mediterranean Diet

One of the goals to which Barcelona aspired in hosting the 1992 Summer Olympics was "to put Barcelona on the map," as mayor Maragall stated (1987). From the beginning, the Olympics were presented as a strategic and urban marketing initiative. The entire planning process and, finally, the success of the event, not only gave the city international recognition but also decisively marked the course of its urban economy and policies. Internationally, the Barcelona's administration was seen as something successful and people began to talk of the city as a model to be followed abroad. Taking advantage of the unexpected international recognition, this expression—the "Barcelona model"—was quickly adopted in the city's promotional policies and by the city government, when Barcelona became a model of urban and architectural improvement in the nineties. Another curious and even more unexpected development occurred in those years, affecting the Catalan dietary identity. That was the period of what came to be known as the "Mediterranean diet" and its promotion by Barcelona, culminating on 16 November 2010, when UNESCO listed it as an Intangible Cultural Heritage of Humanity.[2]

---

2   "The Mediterranean diet constitutes a set of skills, knowledge, practices and traditions ranging from the landscape to the table, including the crops, harvesting, fishing, conservation, processing, preparation and, particularly, consumption of food. The Mediterranean diet is characterized by a nutritional model that has remained constant over time and space, consisting mainly of olive oil, cereals, fresh or dried fruit and vegetables, a moderate amount of fish, dairy and meat, and many condiments and spices, all accompanied by wine or infusions, always respecting beliefs of each community. However, the Mediterranean diet (from the Greek *diaita*, or way of life) encompasses more than just food. It promotes social interaction, since communal meals are the cornerstone of social customs and festive events. It has given rise to a considerable body of knowledge, songs, maxims, tales and legends. The system is rooted in respect for the territory and biodiversity, and ensures the conservation and development of traditional activities and crafts linked to fishing and farming in the Mediterranean communities of which Soria in Spain, Koroni in Greece, Cilento in Italy and Chefchaouen in Morocco are examples. Women play a particularly vital role in the transmission of expertise, as well as knowledge of rituals, traditional gestures and celebrations, and the safeguarding of techniques." *Convention for the Safeguarding of the Intangible Cultural Heritage. Nomination File No. 00394 2010.* Nairobi: UNESCO.

The notion of a Mediterranean diet had a long gestation and a clear promoter, Ancel B. Keys, who correlated the incidence of cardiovascular diseases with high cholesterol levels, an emerging problem among American executives at the time. He proved that the lifestyles and diets on the shores of Mediterranean were the reason for the reduced presence of lipids in people's blood there. Based on these observations, he drew up an "idealized construction," the Mediterranean-style diet, to promote more rational dietary habits in the United States (Keys and Keys 1959).

The scientific debate lasted decades and its effects on the Mediterranean area were delayed for many years, but in the end it became an issue of identity and promotion. It is interesting to find that a major newspaper of high circulation like *La Vanguardia* should make no mention of the "Mediterranean diet" until May 1987. This first rather unscientific reference—and a tangential one at that—came from the Italian actor Ugo Tognazzi during a visit to Catalonia to promote Italian pasta: "North Americans have discovered that the Mediterranean diet is the ideal way of eating and now, in my country, the idea that pasta is fattening is no longer given any credit. The important thing that has been found is that balance is the key." Shortly afterwards, on 16 July, the Spanish Minister of Agriculture was already touting the healthy dietary habits of the Spaniards: "Spain has the highest consumption of fruit, vegetables and vegetable oils in the EEC" (*La Vanguardia* 1987). He also highlighted Spanish shoppers' preference for traditional venues, just as the process of remodeling Barcelona's market halls was getting underway. From this moment on, references to the Mediterranean diet become more common.

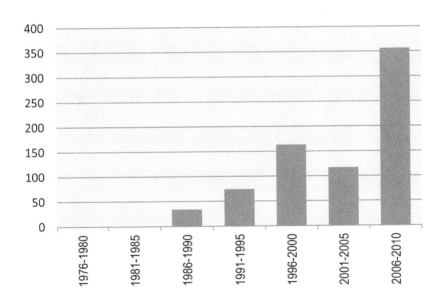

Figure 10.4. Number of references to "Mediterranean diet" in
La Vanguardia

In 1995 the Association for the Advancement of the Mediterranean Diet was founded in Barcelona with the support of local businesses, its mission being to "encourage the consumption of traditional Mediterranean products" (Vallejo 1995). The following year, 1996, saw the creation of the Mediterranean Diet Foundation and the Barcelona Declaration on the Mediterranean Diet, promoted by the FAO (UN Food and Agriculture Organization), the Spanish Ministry of Agriculture and the Barcelona City Council. In February 2005, the Mediterranean Diet Observatory was formed in Barcelona with the aim to encourage UNESCO to declare the Mediterranean diet a form of world heritage. The campaign, in which four countries were involved, was successful in achieving its goal: on November 16, 2010, the Mediterranean diet was listed as an Intangible Cultural Heritage of Humanity.

In the 1995 election campaign for the Barcelona City Hall, the restaurateurs' guild asked Turisme de Barcelona, the Tourism Consortium formed by Barcelona City Council and the Chamber of Commerce, to promote the Mediterranean diet to boost Barcelona's appeal, showing itself to be one of the many groups interested in the campaign (Ricart 1995). In fact, tourism has had a significant growing impact on the city's economy. Suffice it to say that between 1993 and 2007, the number of tourists and overnights in Barcelona hotels grew by 190 per cent and 220 per cent, respectively (*Activitat turística* 2008). The appeal and major culinary potential of Barcelona have become an increasingly important factor in the city's tourism campaigns since the 1990s, activating institutional supports and initiatives in this way. One example of this was the Gastronomy Year, running from March 2005 to March 2006, with the stated aim of taking advantage of recognized Catalan creative talent and the tradition of Mediterranean cuisine, as announced by the second deputy mayor and vice-president of *Turisme de Barcelona*, Jordi Portabella, who was also responsible for the renewal of the market halls (*La Vanguardia* 2005). It was considered that cuisine was one of the main attractions strengthening Barcelona as a cultural tourism destination, so the same successful strategy used previously to promote the link between Barcelona and the works of the architect Antoni Gaudí in 2003, and between Barcelona and the painter Salvador Dalí in 2004, was applied once again.

As a result, in the last two decades a very creative signature cuisine has arisen in Catalonia which has brought many chefs to the forefront of public life and, in some cases, made them international celebrities. The foremost and most highly acknowledged of all is unquestionably Ferran Adrià, the only chef to have been invited to take part in an art exhibition of international prestige as was *documenta 12*, in Kassel, Germany in 2007. These world-renowned chefs have actively contributed to the promotion of Catalan markets and cuisine. The so-called traditional gastronomy of Catalonia, already clearly bolstered over the previous decades, was promoted and systematized even more through the actions of the Catalan Culinary Institute, which in 2002 believed it necessary to enhance the Catalan Cuisine brand, as is stated in the prologue of the *Corpus de la Cuina Catalana* published in 2006, the first inventory of recipes from the popular culinary heritage of Catalonia. In 2011 it published the more ambitious *Corpus del patrimoni culinari català* and, after the successful bid of the Mediterranean diet, it launched the bid of Catalan cuisine as an Intangible Cultural Heritage of Humanity.

Multiple Inclusive Identities

Although the act of defining heritage may be construed as a powerful instrument of ideology, control and social engineering—which creates, to a greater or lesser extent, latent dissonances, exclusions and conflicts (Ashworth 2008)—at the same time each of us participates in multiple identities and this multiplicity enhances heritage's inclusive nature (Risse 2003). Cuisine and "edible" identities are thus used as a device to integrate immigrants while also illustrating the complexity and intricate social, political and cultural resonances that this process acquires in the case of Barcelona and Catalonia.

Public awareness of the vigorous growth of immigration in Barcelona over the last two decades has increased, especially since 2000, for two main reasons: firstly, the increased presence and visibility of groups belonging to cultures that were not common in the city until recently—such as North Africans, sub-Saharan Africans, Asians or Latin Americans—and secondly, the spread of xenophobic attitudes observed in other European countries with a more established immigration history, which appear to provide an insight into developments that could take place here in the future (Paglai 2008). The awareness that cultural unfamiliarity and a lack of interaction generate fear, incomprehension and potential conflicts has given rise to public actions aimed to integrate recent immigrants and promoted initiatives such as intercultural festivals where music and cuisine, among other shared interests, always play a fundamental role. In this respect, mention should be made of *Karakia*, a program broadcast on Catalan television, because it focuses on presenting the cuisine of newcomers from all over the world as a sign of an identity and a heritage to be shared with everyone and especially with locals. Even more significant is the longevity of this program, as it was first broadcast in 2001 and still appears weekly (http://blogs.tv3.cat/karakia).

The first program in this series presented the scope of the migratory process and dealt with the phenomenon from the friendlier angle of interculturality—an idea related to the contemporary understanding by locals of multiculturalism as something mainly inclusive. Since that first show, it has continued to present the more domestic facet of the culinary process as an expression of popular culture and a shared act. It has attempted to bridge differences between cultures and show the changing reality of our country. Generally, the immigrants who take part in the program are characterized by a notable degree of integration. They often speak Catalan and reflect a conflict-free coexistence, offering details on their origins, their culinary traditions and the composition and preparation of various dishes. The program enters their homes, participates in the process of preparing the food and acts as one more guests at the meal, thereby accentuating the ritual of exchange and hospitality. The triple revival of traditional Catalan cuisine, the Mediterranean diet and the culinary traditions of the newcomers to Catalonia overlay one another as perfectly compatible and shared identities, representing both the permanence of the past and the building of the present. One could say that food here is a vehicle for cultural coexistence rather than for ideological boundary-making, like something neutral that everyone can share and learn.

This multiple inclusive dimension has also proved important in the case of municipal market halls. The intense new immigration has had a twofold impact. On the one hand, changes in consumption habits are evidenced by the fact that the market halls, prompted by the globalization process, add new products to their offering, and also by the fact that shoppers are modifying their shopping habits and purchasing traditional products, as may be seen from the notable increase in the consumption of butcher's off-cuts and offal, which had lost much of their previous market share with the improvement of the economy and the dietary changes of recent years (Contreras 2007). On the other hand, many of the newcomers have joined the ranks of the market stall-keepers. Indeed, the market halls present a broad spectrum of jobs that do not require special qualifications or education, as well an environment that is open to affordable business initiatives, especially for those who are familiar with the consumption habits of the new consumers. These changes and the integrating nature of the market halls have been the promotional aims of the Municipal Institute of Markets. In fact, in 2007 it held a European Seminar on Immigration and Commerce with the subtitle "Commerce and Social Cohesion," promoted by the key official Catalan bodies and community organizations. The conference by Marçal Tarragó, author of the PECAB in the 1980s, demonstrates the consistency and continuity of the urban policy guiding these concepts. It also underlines the wide consensus on the essential role of urban commerce and local culture as fundamental factors in the model of the compact, complex cohesive city that Barcelona wishes to be. The term commercial fabric is understood as a tool of social inclusion because it opens new opportunities in a world of commerce that has been considered a social elevator over these last few years and because it has given visibility to our cultural diversity and its manifestations as the fundamental pillars of social cohesion. It also highlights the integrating capacity that stems from its cultural reciprocity, and from the fact that it has activated interaction, networking and a new approach channeled through associations.

Interwoven Processes of Heritage Reinvention

Although they have not been admitted to legally envisaged preservation mechanisms, markets halls, food and cuisine have continued to enter the ledgers of heritage and the defense of clearly identity-related traits. They are too alive, too open and too interwoven into the fabric of everyday life, however, to be suitable for a "museumification" process, which severs an object's ties by transplanting it from its symbolic order "on to an order of history, science and museums" (Baudrillard 1983). Public market halls, food and cuisine are not elitist by nature, although they frequently become vehicles for class-based markers of taste and distinction (Bourdieu 1979). They are not static objects but processes. Culinary identities and the reality of the market hall itself can only be understood as processes, living and basically intangible, forever under construction. (Ashworth 2008). They cannot be reduced to a single reality, a concluded narration or a closed doctrinal body. They are multiple polymorphic realities with a great plasticity. Their study clearly demonstrates the eminently relational, active and creative nature of these processes, allowing one to explore their emergence, consolidation, changes and possibly even their apparent incongruence, which are factors confirming the various vectors that activate their discovery and reinvention (Cosgrove 2003). For example, despite its ductile integrating nature and its rejection of mere preservation, the "Mediterranean diet" has become an immaterial heritage of humanity. Likewise, it may be noted that the reinvention of this apparently malleable and ephemeral edible heritage has coincided with the remodeling and reinvention of the market halls, turning them into a network of authentic heritage landmarks of the city.

The revival of Catalan cuisine and the market halls was initially associated with a resistance to imported lifestyles, which had already demonstrated their negative effects in other European countries, such as the loss of culinary quality and the destruction of the commercial fabric that formed the backbone of neighborhoods. In the process that has been analyzed here, the underlying issue from the start was the preservation of traditional ways of life, which involved the maintenance of the inherent logic of proximity. While it is true that the expansion of traditional cuisine has involved the recovery of some local crops that might otherwise have been abandoned, if one considers the general evolution of things it will be found that the last few decades give a totally different view in sharp contrast to this bucolic image of restoration. If the majority of the fruit and vegetable stall-keepers in 1849 brought in their products from a radius of no more than 20km (Saurí and Matas 1849), by 1994 the mean distance of the fruit and vegetables reaching the central wholesale market was 472km (*Informació estadística* 2005). This distance has grown even more abruptly in more recent times, to such an extent that by 2005 the mean distance was 1,748km (*Informació estadística* 2005). This observation shows that we have not restored the logic of the past in any way, but rather that some assets of the past have reinvented themselves to take root in a completely different world, serving to build a new reality and a renewed identity.

References

Activitat Turística 1993–2007. 2008. Barcelona: Ajuntament de Barcelona. [Online]. Available at: http://www.turisme2015bcn.cat/files/7931-4-arxiuCAT/ ACTIVITAT%20TURISTICA%201993%202007.pdf (Accessed 4 July 2012).

Appadurai, A. 1996. *Modernity at Large. Cultural Dimensions of Globalization.* University of Minnesota Press, 178–82

Arribas, L.E. and Van de Ven, J. 2003. Políticas sectoriales adaptadas e insuficiencia analítica: la regulación del comercio minorista. *Quaderns de Política Econòmica. Revista electrònica.* 5, 2–11. http://www.uv.es/poleco/ (Accessed 4 July 2012).

Ashworth, G.J. 2008. Heritage: Definitions, Delusions and Dissonances, in *Heritage 2008. World Heritage and Sustainable Development,* edited by R. Almoêda et al. Lisbon: Greenlines Institute, 3–9.

Baudrillard, J. 1983, *Simulations,* Columbia University New York City. [Online]. Available at: http://emilylutzker.com/enlightenment/art_media/Baudrillard_sim.pdf (Accessed 22 September 2012).

Bocuse, P. 1976. *La cuisine du marché.* Paris: Flammarion.

Bourdieu, P. 1979. *La Distinction. Critique sociale du jugement.* Paris: Éditions de Minuit.

Cervellati, P.L., Scannavini, R. 1976. *Bolonia: política y metodología de la restauración de centros históricos.* Barcelona: Gustavo Gili.

Contreras, J., 1992, Alimentación y cultura: reflexiones desde la Antropología, in *Revista Chilena de Antropología, n. 11.* Santiago de Chile, Universidad de Santiago de Chile, 95–111

Contreras, J., 2007, Immigració i pràctiques alimentàries. *TECA [Associació Catalana de Ciències de l'Alimentació], 11,* 26–34

Convention for the Safeguarding of the Intangible Cultural Heritage. Nomination File No. 00394 2010. 2010. Nairobi: UNESCO.

Corpus de la Cuina Catalana. 2006. Barcelona: Columna Edicions.

Corpus del patrimoni culinari català. Institut Català de la Cuina. 2011. Barcelona: Edicions de la Magrana.

Cosgrove, D. 2003. Heritage and History: A Venetian Geography Lesson, in *Rethinking Heritage. Culture and Politics in Europe,* edited by R. Shannan Peckham. New York: I.B. Tauris, 113–23.

Di Giovine, M. 2009. *The Heritage-Scape. UNESCO, World Heritage and Tourism.* Lanham, MD: Lexington Books.

Duncan, B. 2010. Catedrales de la dieta mediterránea. *La Vanguardia (Monográfico Especial: Mercados),* 18 September, 1–2.

Featherstone, M. 1991. *Cultura de consumo y posmodernismo,* Buenos Aires: Amorrortu editores.

Guàrdia, M., Oyón, J.L. (ed.) 2010. *Hacer ciudad a través de los mercados. Europa, siglos XIX y XX.* Barcelona: MUHBA, Museu d'Història de la Ciutat de Barcelona.

Harvey, D. 2002. The Art of Rent: Globalization, Monopoly and the Commodification of Culture, in *A World of Contradictions. Socialist,* edited by Panitch and Leys Register, 93–110 http://www.yorku.ca/socreg/ (Accessed 4 July 2012).

Informació estadística del mercat central de fruites i hortalisses de MercaBarna. 2005. Barcelona: MercaBarna.

Keys, A., Keys, M. 1959. *Eat Well and Stay Well.* New York: Doubleday and Company, Inc.

La Vanguardia. 1987. España tiene el consumo de frutas, hortalizas y aceites vegetales más alto dentro de la CEE. *La Vanguardia,* 16 July, 22.

La Vanguardia. 2005. La alta cocina es la última oferta de la promoción urbana. *La Vanguardia (Vivir),* 31 December, 2.

López de Ayala, M.C. 2004. El análisis sociológico del consumo: una revisión histórica de sus desarrollos teóricos, in *Sociológica/5.* http://ruc.udc.es/dspace/bitstream/2183/2725/1/SO-5-6.pdf (Accessed 4 July 2012).

Luján, N., Perucho, J. 1970. *El libro de la cocina española.* Barcelona: Danae.

Luján, N. 1993. *Vint segles de cuina a Barcelona.* Barcelona: Ediciones Folio, S.A.

Luján, N. 1977. Prologue, in Vázquez Montalbán, A. 2004. *L'art del menjar a Catalunya.* Barcelona: Edicions 62, 11–18.

Maragall, P. 1987. *Per Barcelona,* Barcelona: Edicions 62.

Memòria, 1987–1991. 1991. Barcelona: Àrea de Proveïments i Consum, de l'Ajuntament de Barcelona.

Pagliai, V. 2008. *The Unmarking of Racist Discourse in Italy and the Role of the Media.* San Francisco: American Anthropological Association http://www.aiserarchive.com/ValentinaPagliai/papers.htm (Accessed 4 July 2012).

Perucho, J. 1999. *Gastronomia i cultura.* Tarragona: El Mèdol.

Pla, J. 1972. *El que hem menjat.* Barcelona: Destino.

PECAB (Pla Especial d'Equipament Comercial Alimentari de la Ciutat de Barcelona). 1990. Barcelona: Ajuntament de Barcelona, Àrea de Proveïments i Consum.

Pla estratègic de Turisme de la Ciutat de Barcelona 2015. Anàlisi Estratègica. 2009. Barcelona: Ajuntament de Barcelona.

Risse, T. 2003. European Identity and the Heritage of National Cultures, in *Rethinking Heritage. Culture and Politics in Europe,* edited by R. Shannan Peckham. New York: I.B. Tauris, 74–89.

Ricart, M. 1995. Maragall insiste en que Roca corteja al PP. El candidato socialista se concentra en los barrios obreros y apela al electorado de centro. *La Vanguardia,* 25 May, 18.

Sandoval, J. 1987. Tognazzi, en Sant Pol: el actor que mejor cuece, fuera y dentro de la cocina la "pasta gansa." *La Vanguardia,* 15 May, 30.

Santamaria, S. 1998. La armonía defendible in *El color en la alimentación mediterránea. Elementos sensoriales y culturales de la nutrición.* Barcelona: Institut Català de la Mediterrània, 223–7.

Santamaria, S. 2005. Prologue, in Pla, J., *El que hem menjat.* Barcelona: Destino.

Saurí, M., Matas, J., 1849. *Manual histórico-topográfico estadístico administrativo (ó sea Guía general de Barcelona).* Barcelona: Imprenta y Librería de D. Manuel Saurí.

Schmiechen, J., Kenneth, C. 1999. *The British Market Hall. A Social and Architectural History.* New Haven, London: Yale University Press.

Vallejo, E. 1995. Buena mesa. Barcelona, sede del Centro Europeo de Alimentación Mediterránea. *La Vanguardia,* 16 March, 84.

Vázquez Montalbán, A. 1977. *L'art del menjar a Catalunya.* Barcelona: Edicions 62.

Vázquez Montalbán, A. 1984. *L'art del menjar a Catalunya. Crònica de la resistència dels senyals d'identitat gastronòmica catalana.* Barcelona: Edicions 62.

Vázquez Montalbán, A. 2004. *L'art del menjar a Catalunya. El llibre roig de la identitat gastronòmica catalana.* Barcelona: Empúries S.L.

Watson J.L., Caldwell M.L. (eds) 2005 *The Cultural Politics of Food and Eating: A Reader*, Malden, MA: Blackwell Pub.

Chapter 11

French Chocolate as Intangible Cultural Heritage

Susan Terrio

Introduction

In 2008 the Parisian leaders of the Confederation of artisanal chocolate and sugar candy producers (*Confédération des Chocolatiers et Confiseurs de France*) got wind of the state plan to seek the registration of the French gastronomic meal on the UNESCO list of intangible cultural heritage forms. Writing to the confederation members, Guy Urbain, the editor of the trade journal with 50 years of experience in the profession, described the Confederation's attempt to include chocolate in the dossier on the gastronomic meal being prepared by state officials for UNESCO consideration. Urbain and the president of the Académie Française du Chocolat et de la Confiserie made the artisans' case to President Nicholas Sarkozy's advisor, Christophe Malevin. Urbain explained that "the goal was to recognize the quality, originality and excellence of artisanal products" and to emphasize the "global interest" in French chocolates. He added that the recent official launch of *chocolats à la française* made it "the perfect time to affirm French candies as an integral part of French gastronomy" (Urbain 2008b: 21). President Sarkozy's advisor received the Confederation members at the Elysée palace and encouraged their efforts, suggesting that sweet gastronomic products could potentially be included in the state dossier then under preparation (Urbain 2008d). A year and a half later, in 2010, when the French gastronomic meal was officially registered on the UNESCO representative list of intangible cultural heritage, there was no specific mention of *chocolats à la française* (UNESCO 2010). This was an interesting omission since the description of the gastronomic meal on the UNESCO web site closely parallels the social rules governing French chocolate consumption in a rapidly growing literature devoted to it. Like the gastronomic meal, chocolate is tied to longstanding confectionery traditions, a growing repertoire of haute cuisine recipes, and social rituals that celebrate important rites of passage. Significant elements include the pleasure of taste, the emphasis on quality, the aesthetic presentation of candies, the design of confectionery art, the importance of gustatory norms, the training of future practitioners as well as the role of chocolate gastronomes and producers with deep knowledge about taste and production.

Despite the official omission of chocolate as a constituent element of French gastronomy, I focus here on the strategic appropriation of the concept of intangible cultural heritage by a community of French artisanal chocolatiers as a devise to define, position, and represent their products, skills, associations, and modes of production in national and global markets. The attempt to have French chocolates recognized as gastronomic heritage by state representatives, expert specialists, and the French public speaks to both the ambiguities and possibilities contained within the 2003 UNESCO *Convention for the Safeguarding of Intangible Cultural Heritage* (UNESCO 2003).

Intangible Cultural Heritage

The convention on intangible cultural heritage that was ratified by 30 nation-states in 2003 created a new category that moves away from the material culture of objects to focus on the living cultural practices and social events that form the basis of group identity and collective memory. The Convention specifically recognizes language, arts, rituals, knowledge, and artisanal skills that are closely tied to communities and that are continuously recreated through time. In this formulation, intangible cultural heritage must be protected because it is dynamic and viable, not salvaged because it is marginal and in danger of extinction. This emphasis on process is closely tied to the agency and identity of community actors who, according to the 2003 convention, select and rework the elements of culture that constitute their own unique heritage (UNESCO 2003).

The UNESCO inventory of heritage lists has attracted global attention and increasing numbers of candidates for inclusion. Nonetheless, it begs important questions related to the politics of culture and the authenticity of heritage. How are social actors in communities to interpret the mandate to safeguard existing cultural heritage? When and how do they start the process of patrimonialization (Bortolotto 2007: 2010)? How does this process affect the form and content of the traditions that are deemed legitimate? Does it center on actual cultural practices regardless of their current vitality and historical pedigree or does it rely on objectified representations of these same practices (Kirshenblatt-Gimblett 2004, Nas 2002, Noyes 2006)? In France, the claims surrounding culture and chocolate necessarily involve complex negotiations with powerful state representatives who have a vested interest in defining and promoting what constitutes national patrimony and gastronomic tradition. The promotion of French chocolates occurs within a global hierarchy of value involving national cultural ideologies (Herzfeld 2004) and handcrafted foods that can be commodified and sold on global markets as both ancient heritage and living practice. As we shall see, French chocolatiers were as concerned with seeking recognition for a proven confectionery heritage as they were in producing new chocolate traditions that they represented as quintessentially French. In the twenty-first century, chocolatiers focus on the transmission of changing repertoires of skill, the organization of highly competitive expositions that display the work of master chocolatiers and confectionery artists, the cultivation of single-origin chocolates that exemplify distinctive local *terroirs*, and the codification of specialized knowledge surrounding taste by community-appointed experts.

French Chocolate, 1988–2000

In 1988 when I began research on artisanal chocolatiers, French consumers patronized family businesses whose products included pastries, chocolates and sugar candies. They purchased mainly milk chocolates, chose from fewer house specialties, saw virtually no confectionery art, and had no specialized chocolate guides. Moreover, what counted as French taste in chocolate was not at all clear. Although a majority of French people claimed to like it, French consumption of chocolate was low compared to other Europeans and it remained strongly associated with gifts exchanged at seasonal, ceremonial, and social occasions. The French tended to eat chocolate at widely spaced intervals and its consumption was constrained by prescriptive rules. Although aficionados have always claimed dark chocolate as the superior taste standard, among the French public there existed comparatively little knowledge or evaluative criteria to differentiate meaningfully among different chocolates. Similarly, artisanal chocolatiers were not recognized as an autonomous craft separate from pastry making and they had no specialized credentials to offer would-be artisans.

To make matters worse, French chocolate producers faced the challenge of intensified international competition from Swiss, American, and Belgian firms in the 1980s. At the same time, European Union representatives proposed reforms that threatened to undermine French norms of production by redefining the composition of chocolate to allow vegetable substitutes for the cocoa butter naturally found in chocolate. Worse, foreign franchise outlets targeted the market for handmade gifts by selling industrial candies in storefronts that closely resembled French craft businesses. Franchise outlets cost a fraction of a fully equipped workshop, retailed their candies for one half to one third the price of handcrafted French candies, and appropriated French merchandising techniques by using elaborate packaging and assigning evocative names to individual candies. Between 1983 and 1989 these foreign outlets captured 48 per cent of the confectionery gift market (Mathieu 1990).

In response, Parisian craft leaders, local chocolatiers, and the state officials to whom they appealed collaborated to develop an appropriate model with which to authenticate French artisans, their goods, and their modes of production. State officials were receptive to their appeals in part because state policies regarding artisanship had shifted since the 1970s. Legislative reforms aimed to preserve artisanal training models and certification procedures. In 1980 the state Office of Ethnological Patrimony began a research program that was intended "to document cultural remnants of traditional France" (Programme savoir-faire et techniques, dossier 1, 1988). In 1989 the Ministry of Culture created the National Council of Culinary Arts with the goal of safeguarding French gastronomy. Chocolatiers institutionalized a separate craft identify by organizing training programs at all skill levels and by offering new state diplomas. In 1990 the Ministry of Education approved chocolatiers' request to organize the first Best Craftsman of France award in chocolate and sugar candy production. Over this period chocolatiers also codified and disseminated a new esoteric taste standard borrowed from wine connoisseurship that was based on dark chocolate. Finally, they turned to the past as a means of authenticating their collective identity and craft culture. They selectively appropriated and reworked a number of different symbols, rituals, and histories-craft, Aztec, local, and national-to affirm their collective identity and to empower themselves in a competitive struggle for legitimacy (Terrio 2000).

By 2000 French chocolate had been transformed from a traditional product sold primarily as a gift to a good newly reinvested with gourmet cachet and cultural authenticity. Chocolate appears in new culturally relevant categories for personal consumption, ranking in connoisseur tastings, and purchase as confectionery art that is commissioned for private sale or entered in prestigious contests. Discerning consumers now choose luxury bars with high percentages of dark chocolate and differentiate among bean varietals and cocoa vintages. This is a trend that began in France and was exported to the U.S. By 2008, French consumers ranked third in chocolate consumption after Great Britain and Germany but ahead of Italy and Switzerland (Urbain 2008e). Chocolate candies were the most popular item and consumers registered a growing interest in high-end, organic and fair trade products as well as candies with less sugar and high percentages of cocoa solids (Urbain 2008b: 25–6). Every year new events are organized in which chocolate is displayed as an object of extravagant spectacle, artistic prowess, and culinary refinement. In the 1990s one of the newest additions to the Parisian exhibits of consumer goods was the *Salon du Chocolat*. When it opened in 1995, the Salon drew 40,000 visitors who witnessed the spectacle of chocolate being melted, tempered, hand-dipped, sculpted, molded, and even modeled in an haute couture fashion review. The chocolate fashion show continues as does the Festival of Gourmet Art which features chocolate jewelry, corsets, gowns as well as paintings, both originals and reproductions, in white, dark, and colored chocolate. In 1998 craft leaders created the Académie française du chocolat et de la confiserie. Explicitly modeled on the literary institution inaugurated in the seventeenth century to codify proper linguistic usage, the 40 chocolate immortals have undertaken the weighty task of writing the first official confectionery dictionary and of disseminating new knowledge related to cacao bean varietals and regional *terroir*.

Chocolatier Skill as Intangible Cultural Heritage

In May 2011, French chocolatiers held the final round of the Best Craftsman of France (BCF) (*Meilleur Ouvrier de France*) contest, the most prestigious state award for artisans. This was the 6th contest organized since 1990, when craft leaders successfully staged the first competition for chocolate and sugar separate from pastry. The contest is a crucially important "tournament of value" (Appadurai 1986: 21) involving competitive, even agonistic, displays of skill which affirm an autonomous identity for chocolatiers, promote the power and integrity of French gastronomy, and authenticate the contestants as both superior artisans and confectionery artists. The contest regulations, the selection of judges, the number of candidates, and the winners reveal much about craft, community, and the constantly evolving practices and knowledge related to chocolate.

The BCF contest is one of a number of national and international competitions in which chocolatiers seek a prominent position within a national hierarchy of craft practitioners. The contest organizers, mainly powerful Parisians, and the contestants see the acquisition of craft skills as inseparable from the communities of practice in which they are learned. Advancement within the community ranks is accomplished through rite of passage tests that separate novice apprentices from accomplished journeymen and, finally, master artisans. Community membership involves learning that combines observation and practice. It is a long-term process whereby a repertoire of collective dispositions, involving precision hand and arm movements, stance, rhythm, and muscle memory are imprinted through guided practice and repetition. Artisans develop both empirical mastery and cognitive understanding of the raw material. Yet competing requires years of practice, self-discipline, mental concentration, physical stamina, and the ambition needed to sustain grueling seven day-a-week routines involving constant trial and error. The 2003 winner, Frank Kestener, explains:

> In school I loved design and the practice came naturally. I always did artistic pieces and specialty candies for competitions. It was training and the opportunity to see what others do. My father is a pastry maker and he was my most important coach. He was part of the organization Gourmet Tradition so he took me to seminars. Later he introduced me to two chocolatiers, Serge Granger and André Rosset who were preparing for the BCF contest. It was their second attempt after they failed the first time [in 1990]. I tried for the title after winning the International Grand Prix in 2002. That was an important test because it proved that I was ready. I had experience. I knew how to discipline myself, to manage my stress, to handle failure, to be well prepared, and to go to the limit of my endurance without losing control. It's a question of character and mental attitude. It wasn't like that for me in the beginning. I may appear calm but I get emotional. Before if things didn't go the way I wanted, I would get angry and break the thing I was making and then disappear for two days.

This demand for endurance is tested in the final round where contestants labor for 27 hours over three days. They work behind closed doors in contiguous work stations while judges circulate continuously requesting explanations and evaluating their technique, technology, organization of work space, management of allotted time and performative presence. Contest juries, dominated by powerful practitioners and gastronomes, judge results in three categories: aesthetic presentation, taste, and work technique. They award scores for entries that include individual candies, molded figurines, and confectionery art. To win the contestant must demonstrate control of the bases of craft as well as the superior skill required to advance it through an innovative use of tools, design and/or taste. The competition is always fierce. In 2011 only one of the four finalists took his place in the pantheon of 18 BCF in chocolate.

The 2011 winner, Frederic Hawecker, attributed his selection to the training he received from current title holders, "whose passion and determination" inspired him to attain their level of excellence (Urbain 2011b: 12–13). His interview echoed the narratives of the craft leaders such as Confederation president Madeleine Lombard who evoked a shared identity grounded in meaningful work as well as the "traditions of the masters of old who worked, suffered, doubted, seeking relentlessly the originality and the aesthetic implicit in new creations" (BCF Award Ceremony 1990). Discursive narratives which focus on skill, ethics, and perseverance suggest a democratic process that is open to all hardworking artisans. In reality, the high standards, time investment, financial resources, and social capital required for success mean that those best positioned to win are artisans born into craft families and skilled through workplace experience. All 19 winners were heirs to or the creators of successful confectionery businesses. Hawecker went through an apprenticeship and worked in a *chocolaterie* he owned with his brother. Hawecker's trajectory corresponded to a craft ideal as he was a *Compagnon du Devoir*. *Compagnonnage* is a state-sanctioned journeyman brotherhood association which is modeled on the guild system of Old Regime France. It is also registered on the UNESCO representative list of intangible cultural heritage for France. Hawecker had spent 7 years working in confectionery houses in six different French cities as well as Brussels. He also worked an entire year assisting title holders at one of the most reputable schools in the food trades, the Ecole Nationale Supérieure at Yssingeaux near Lyon, site of the BCF contest (Urbain 2011b: 12–13).

The gentrification of chocolatier skill resembles the changes within professional cuisine in the late nineteenth century when Parisian chefs used culinary expositions to extend the hierarchy of the arts to produce French haute cuisine (Trubek 2000: 110–27). The intricate set pieces which are the centerpiece of the BCF contest must transcend the standards of the decorative arts to enter the rarified domain of the fine arts. The challenge in the design and execution of confectionery art is to prioritize form and aesthetics over function and pragmatics while still remaining faithful to contest themes; in 2011 it was Christmas. To have the edible masquerade as the fanciful in the form of lifelike Saint Nicholas figurines, sleighs, holly, reindeer, and snow-covered pines is to engage in a higher form of abstract play. At the same time, both original art and candy specialties must respect tradition and exemplify the control of basics related to technique and taste. Title winners must demonstrate the technical competence which allows them to master a highly demanding and unstable raw material such as chocolate.

Hawecker's winning set pieces exceeded expectations. His sculptures soared upward, posing and resolving the challenge of balancing height with structural integrity and design ingenuity. The most intricate piece featured a gift-laden sleigh balanced precariously atop a cacao tree.

Winners join the rarified ranks of masters who are the primary guardians of an artisanal aesthetic and the community gatekeepers. In the 2011 competition the five judges designated to evaluate work technique were all Best Craftsmen of France in chocolate and sugar candy production. Of the 15 judges appointed to judge taste and presentation, six were title holders. What is at issue in this competition is the certification of craft traditions and a unique taste standard as well as the rank, reputation, and potential economic gain for French artisans in national and global markets. In the late 1980s, France's renown as a nation of luxury craft production and culinary arts lent weight to the authentication of French chocolate as national patrimony and living culture. Twenty years later, the global confectionery landscape had evolved considerably. The global prominence of French chocolates and the adoption of taste standards borrowed from wine connoisseurship now reaffirm the cultural authority and esoteric cachet of French gastronomy itself.

French chocolatiers no longer have to leave France to get advanced training in Belgium or at the Swiss Coba school, as did Robert Linxe, the founder of the famous Parisian chocolaterie, La Maison du Chocolat. Moreover, the role played by Linxe in the dissemination of esoteric chocolate taste earned him a Legion of Honor award—the highest recognition given by the French state. Confectionery artisans now increasingly come to France from outside the metropole to train and to compete in prestigious competitions and expositions. For example, two of the 2011 BCF finalists work in Montreal and Las Vegas respectively. When craft competitions are held abroad sponsors frequently turn to French experts to organize the contests. For example, the World Pastry Championship, held every two years in the United States, attracts pastry chefs and chocolatiers from all over the world who use the regulations developed by French artisans.

Chocolats à la française: *Terroir* and Taste

In early 2008, confederation leaders charged members of the Académie Française du Chocolat et de la Confiserie to define *chocolats à la française*. The Académie is an association of powerful producers, scientists, agronomists, academics, and gastronomes. They defined French chocolates as "rich in cocoa solids, made with pure cocoa butter, and as products that exemplify a culture of excellence, creativity, and innovation as well as the brilliance (*rayonnement*) of French artisans" (Urban 2008a, 21). Academicians linked the growing repute of French chocolates abroad to a "culture of the aesthetic;" "to a universal chocolate terminology" inseparable from the French language, an essential medium of gastronomy;" and to "the constant search for different flavors " (Urbain 2008c, 19). Nonetheless, the official definition of French chocolates was intentionally non-specific. It did not specify a minimum percentage of dark chocolate or a regional origin. "The essential thing," Academicians insisted, "is quality, with clearly marked ingredients" (ibid.).

Craft leaders' strategies speak to the continuing challenges of cultivating the taste of a distinctive *terroir* and of authenticating the production of raw materials that are cultivated, harvested and increasingly processed into semi-finished products outside of France, in the major chocolate-producing countries of Africa, Asia and South America. The celebration of French chocolate as cultural patrimony involves a complicated dialogue among gastronomes, producers, consumers, and state representatives. This dialogue entails a strategic manipulation of the gaps in knowledge that intensify as the distance from cultivation to consumption sites increases and as supply chains lengthen. My initial research showed that most French producers had no direct knowledge of the tropical raw material or of small-scale cacao cultivation. Artisanal chocolatiers occupy a specialized niche within a fully industrialized sector and purchase semi-finished chocolate blocks or *couverture* from a few large manufacturers who continue to dominate the market. Although consumption criteria in published guides emphasized the importance of climate, soil, cacao varietals, and estate growths, the vast majority of the French producers I studied learned about cacao-producing regions from training seminars and marketing brochures, not from direct knowledge of the harvest and transformation of raw cacao.

Twenty years later, sourcing high quality, single-origin beans, and classifying estate growths (*les grands crus de terroir*) has become a major preoccupation for gastronomes, for artisanal producers, and for consumers. The confederation, with financial support from industrial manufacturers such as Barry Callebaut, organizes yearly trips to cacao-producing countries so that artisans will gain direct knowledge of the cultivation and harvest of cacao beans. In May 2011, 95 professional chocolatiers, chefs, and pastry makers attended an international conference in the Bahia region of Brazil. A collaboration between international and national research and consumer groups led to the creation of the International Cocoa Award to recognize growers who produce the best quality beans (Urbain 2011a: 8–9). Increasing numbers of manufacturers are self-described *couverturiers* or producers of chocolates from the same *terroir* or region. Others have moved to control the entire production process with beans to bar products. Valrhona, for example, cultivates long-term partnerships with small producers in Venezuela, Trinidad, and Madagascar, and the firm established its own cacao plantation in Venezuela. By providing technical assistance to local growers, Valrhona experts saved a rare and endangered bean varietal, the Porcelana (Urbain 201: 47–9). At the same time, Valrhona and other manufacturers have continued to market luxury chocolate bars by origin and/or varietal and to emphasize the selection of quality beans from specific *terroirs* such as the Alpaco from Ecuador or estate growths (*chocolats de domaines*) such as Palmira from the Valrhona plantation.

More complex knowledge about production demands more detailed knowledge about consumption. Consumers must be able to differentiate the rare craft version of chocolate from the mass-produced one by using new standards to evaluate taste. In this process, gastronomes or taste experts have acquired a new prominence, in particular those who occupy the rarified ranks of the Académie. At the chocolatiers' 2008 conference Academician Katherine Khodorowsky delivered the keynote address and addressed the issue of the chocolate expert (*chocolatologue*). Who corresponds to this description? Certainly "not a young person just out of training" or even a professional "who makes nice chocolates of average quality." Nor could it be the artisan who thinks cocoa and chocolate are the same thing or the business owner who is confused about where chocolate originated. A chocolate expert is a professional "rich in cultural knowledge-in the widest sense-on the cacao producing countries, on the history of chocolate and on the evolution of taste" (Urbain 2008a: 13) A chocolate expert is someone who continues learning and growing throughout his professional life and who transmits his knowledge to future generations.

Conclusion

This chapter has focused on the attempt by one community of French artisanal chocolatiers to identify their products, skills, and craft community as forms of intangible cultural heritage. In contrast to the critics of the 2003 UNESCO convention who argue that the recognition of cultural patrimony ritualizes outdated practices and consigns traditions to the status of living fossils, we have seen how a community of artisanal chocolatiers attempted to have their goods and rituals included as part of the French gastronomic meal. French chocolatiers are as concerned with seeking recognition for a proven confectionery heritage as they are in producing new chocolate traditions that they represented as quintessentially French. Their initiatives draw on essentialist notions of cultural traditions that persist unchanged through time and on constructivist understandings of tradition that suggest evolving identities and practices within dynamic communities of practice. Contemporary chocolatiers' use of craft rituals, associations, and practices show how they continue to rework and represent their own particular heritage.

References

Appadurai, A. (ed.) 1986. *The Social Life of Things*. Cambridge: Cambridge University Press.

Bortolotto, C. 2007. From objects to processes. UNESCO's intangible cultural heritage. *Journal of Museum Ethnography*, 19, 21–33.

_____. 2010. Le trouble du patrimoine culturel immatériel. *Terrain* (Online) Available at: http://terrain.revues.org/14447 (Accessed: 29 November 2011).

Herzfeld, M. 2004. *The Body Impolitic: Artisans and Artifice in the Global Hierarchy of Value.* Chicago: Chicago University Press.

Kirschenblatt-Gimblett, B. 2004. Intangible cultural heritage as a meta-cultural production. *Museum International,* 221–2, 53–66.

Mathieu, J. 1990. *La Confiserie du chocolat: diagnostic de l'univers et recherche d'axes de développement.* Paris: l'Institut d'Observation et de Décision, La Direction de l'Artisanat du Ministère de l'Artisanat et du Commerce.

Nas, Peter J.M. 2002. Masterpieces of oral and intangible culture. Reflections on the UNESCO World heritage List. *Current Anthropology,* 43(1), 139–48.

Noyes, D. 2006. The Judgment of Solomon. Global protections for tradition and the problem of community ownership. *Cultural Analysis* 5, Theory/Policy, 27–55.

Terrio, S.J. 2000. *Crafting the Culture and History of French Chocolate.* Berkeley: University of California Press.

Trubek, A.B. 2000. *Haute Cuisine. How the French Invented the Culinary Profession.* Philadelphia: University of Pennsylvania Press.

UNESCO. 2003. Convention pour la sauvegarde du patrimoine culturel immatériel. Retrieved November 29, 2011 at http://unesdoc.unesco.org/ images/0013/001325/132540f.pdf.

UNESCO. 2010. Convention pour la sauvegarde du patrimoine culturel immatériel. Comité intergouvernemental de sauvegarde du patrimoine culturel immatériel. Dossier de candidature no. 00437 pour l'inscription sur la liste représentative du patrimoine culturel immatériel en 2010. France. Retrieved November 29, 2011 at http://unesdoc.unesco.org/images/0013/001325/132540f.pdf.

Urbain, G. 2007. Une gamme qui a du sens. *Chocolat et Confiserie Magazine* 422, 40–43.

Urbain, G. 2008a. Devenez chocolatologue. *Chocolat et Confiserie Magazine* 424, 12–15.

Urbain, G. 2008b. Les chocolats à la française. *Chocolat et Confiserie Magazine* 425, 21.

Urbain, G. 2008c. Ils partent à la conquête du monde. *Chocolat et Confiserie Magazine* 426, 16–23.

Urbain, G. 2008d. Les chocolats "à la française" à l'Elysée. *Chocolat et Confiserie Magazine* 428, 15.

Urbain, G. 2008e. L'évolution du marché mondial. *Chocolat et Confiserie Magazine* 428, 21–7.

Urbain, G. 2010. Valrohona: une approche différente. *Chocolat et Confiserie Magazine* 441, 47–9.

Urbain, A. 2011a. Champion du monde. *Chocolat et Confiserie Magazine* 442, 59–61.

Urbain, A. 2011b. Frédéric Hawecker. Le nouveau M.O.F. *Chocolat et Confiserie Magazine* 445, 12–13.

Chapter 12

Daily Bread, Global Distinction?
The German Bakers' Craft and Cultural
Value-Enhancement Regimes

Regina F. Bendix

The Lord's Prayer includes the simple wish, "give us this day our daily bread." Indeed, bread made of flour from one or more grains, leavened or unleavened, is one of the most wide-spread basic foods on earth, rivaled at best by rice.[1] So basic and so widespread is bread that one might think its universal significance so self-understood that no mention need be made of it. Thus it was with some surprise that I took note of the following news item reported in the German media on May 22, 2011:

> German Agriculture Minister Ilse Aigner (CSU) wants to have German bread be protected as World [Intangible] Cultural Heritage. 'In its variety and quality, German bread is unique,' Aigner declared on Sunday in Berlin. With this, she supports suggestions made by the German bakery trade federation to create a 'Bread Registry'. (see 'Vorschlag' 2011).

As I soon realized, this was not the first press report on the topic of "German bread as cultural heritage." Already on March 21 of the same year, the *Zentralverband des deutschen Bäckerhandwerks* (German Federation of the Bakery Trade) had placed the following notice on its homepage:

> With its globally unique diversity, Germany is rightly regarded as 'the land of bread.' But how many varieties actually exist? The answer will come soon

1   Thanks go to my student assistant Nora Kühnert who assisted in the research for this chapter, Dr. Sven Mißling, postdoc in international law who assisted with retrieving documents from German parliamentary discussions, as well as to other colleagues and junior researchers in the Göttingen research group on the constitution of cultural property, supported by the German Research Foundation since 2008 (DFG FOR772) and colleagues within the Göttingen Center for Modern Humanities (ZTMK). In addition to thanking the DFG for supporting our research endeavors, I also thank the Göttingen Lichtenberg Kolleg which granted me an association during 2011–12, during which the chapter was written. All translations from German are my own.

thanks to the Bread Registry the German *Zentralverband des Deutschen Bäckerhandwerks* is now compiling. For the first time, this abundance will be documented. And that is just the beginning, since by 2012 German bread is to become UNESCO World Cultural Heritage![2]

The bakers' trade federation is the late modern institution representing what in early modernity were the bakers' guilds. They put out four more press releases providing information about how far the registry had progressed. When I asked how the efforts to nominate varieties were being implemented, I received the following answer from the office responsible for foodstuffs law at this federation:

Dear Mrs. Bendix,

Together with the Academy of the German Bakers' Craft in Weinheim (ADB Weinheim) and the Institute for the Quality Assurance of Baked Goods (IQBack e.V.), the *Zentralverband des Deutschen Bäckerhandwerks* is creating the so-called Bread Registry (www.brotregister.de). On behalf of its members, the *Zentralverband* is applying to be taken onto the Intangible Cultural Heritage list. (Email received September 30, 2011)

A visit a few days later to this registry site revealed that an impressive 2,509 types of bread had thus far been registered. Alas, any heritage nomination requires that a state ratify the corresponding international convention, which Germany has thus far not done. I sent a query to the head of the Culture Bureau at the German UNESCO Commission in Bonn in early October 2011 and received the response that there was no discussion at the German UNESCO Commission about a "German Bread Registry."[3] At the time of this writing, July 2012, the German parliament has not yet ratified the 2003 UNESCO Convention on World Immaterial Cultural Heritage. Thus while the number of breads registered on the German Bread Registry continues to rise, the biggest hurdle German bakers face with regard to a UNESCO listing lies within the political process in Germany itself.

From this cluster of press reports, inquiries, and facts on a potential bread-nomination, a number of research questions arise about the contemporary processes surrounding the "valuation of culture," which I wish to address in turn. The goals pursued by the various actors involved are worthy of serious consideration, as they provide insight into what has been set loose by a now 40-year-old "World Cultural Heritage" regime and parallel value-added procedures. To shed light on the German bakers' actions, I will examine why these actors develop the ambition to render a basic foodstuff into world

---

2   Online press release 009/2011; accessed November 15, 2011.

3   Thanks for this inquiry to my student Sina Wohlgemuth who carried out a practicum with the culture division of the German UNESCO office, and communicated this information to me on September 28, 2011, via email.

cultural heritage. I will also study the role the state has been allotted within UNESCO's heritage conventions and follow with an examination of the role that ethnological and anthropological fields of inquiry have played—or not played—in this particular heritage-making endeavor. Finally, I will probe the motivation of the German bakers more deeply and lay open the rationale for pursuing a UNESCO nomination for intangible cultural heritage for German breads rather than pushing for any other extant value-added regime.

A Basic Foodstuff as Cultural Heritage?

Thinking about bread, one would logically first think about its taste and texture, variety and history. The German star cook Wolfgang Siebeck did, when he addressed the potential heritage status of German bread in his column in a German weekly. Not interested in the political hurdles or the fine points of the UNESCO heritage regime, he immersed himself immediately in the history of German bread, and eventually argued that the only really tasty bread, though it does get rock-hard, is found in the Alps. His conclusion was that "if it is a matter of originality, then (only) pumpernickel and the related black bread [*Schwarzbrot*] have a chance at World Heritage status." Siebeck, the cook, naturally and rightfully is focusing on taste. However, in the context of cultural heritage, we can leave aside the question of which bread, or breads, or whether all German bread (for whatever taste-related reasons) really deserves the status of being called an intangible world heritage. The primary issue here is not bread itself, and not the cultural practices of sowing, reaping, grinding, mixing, kneading, and baking that surround it. It is not even about the pleasure taken at eating good bread. Rather, it is about the practices that lead to valuing this cultural product.

Barbara Kirshenblatt-Gimblett has called the heritage-labeling of bits of culture a "'value-added' industry" (Kirshenblatt-Gimblett 1995: 370–72). She coined the term "metacultural production" for this process which includes attitudes, values about traditional cultural expressions,[4] and their instrumentalization (Kirshenblatt-Gimblett 2004). It has been 16 years since she published her five theses on heritage (1995), and her argumentation at the time certainly constituted a watershed in both tourism and heritage research. In light of the growing density of heritage designations and the space they take on the economic and political stage, I would, however, argue that these processes should not be called "meta" any longer. They are themselves cultural practices, and if one looks at some existing or planned world heritage sites and intangible cultural practices in creating officially acknowledged cultural heritage, these processes seem to have almost become more significant than

---

4   "Traditional cultural expressions" (TCEs) and "traditional knowledge" (TK) are the technical term used within the World Intellectual Property Organization (WIPO) to refer to practices and knowledges designated as folklore within an American context.

the slices of intangible cultural heritage, the existing and designated cultural monuments, or the cultural heritage landscapes themselves.

It is not that surprising that foodstuffs, or dining complexes, are honored as immaterial cultural goods. Individual foods as well as cuisines have a long history as distinguishing (as well as derogatory!) markers of nation and ethnicity. UNESCO's Intangible Cultural Heritage (ICH) convention affords an opportunity to enhance the symbolic value of cookery on the world stage—with recipe knowledge and cooking crafts as intangible, passed-on traditions. After a first wave of nominations from the realm of customs, musical styles and narrative forms to become ICH, nominating stakeholders from around the world have gradually widened the scope of traditional practices to be considered. At the fifth meeting in Nairobi in November of 2010, for example, UNESCO's Intergovernmental Commission for selecting immaterial cultural goods decided to include Mediterranean cuisine as world cultural heritage. Greece, Spain, Italy and Morocco, working together, were successful in this—an unusual but increasingly common practice, as the Italian ethnologist Alessandra Broccolini, who was involved in this application, has documented (Broccolini 2013). In the same year, Croatia was successful in having its gingerbread be placed on the list. Mexico, too, convinced the selection committee in the area of nutrition with a nomination that even met the precondition demanded by UNESCO, namely that a tradition must be endangered and that a place on the world heritage list would help preserve the tradition or might even lead to its revival. France, with a certain amount of cockiness and clever packaging of its claim, was able to have the "French meal" be nominated (Tornatore 2013). So it is no surprise that stakeholders in nations like Korea and Japan were following suit, or at least pursued the worthiness of their respective meal cultures within the media though thus far no nominations have reached committee decision ("Panel" 2011, "Korea.net" 2011).

In light of this competition as well as the linkages into the world nutritional discourse occasioned by food-nominations of ICH,[5] it might be understandable that German bakers, too, want to be on the heritage map, though, as will be shown, this is not the only and not the major reason to pursue such a goal. But the problem remains that Germany has yet to ratify the 2003 ICH Convention.

What Leads a German Federal Minister to Support a Potential Heritage Nomination for Bread?

Bread is ultimately an agricultural product, and Ilse Aigner as German minister of agriculture is thus well advised to examine and support potential opportunities

---

5   The World Public Health Nutrition Organization (2011) announced the 2010 ICH nominations and linked them into world discourses on ecologically sound agriculture and the nutritional value of traditional cuisine in comparison to fast food.

that might increase the value of a German product within and beyond the realm of German consumption. One may nonetheless wonder why Aigner—as well as the German bakers' association—turn to the heritage route. World heritage is a global label. Agreements under international law, as made by every sub-organization of the UN including UNESCO, do not have the force of law. The UN is composed of member states that (still) have a territorial basis,[6] and every agreement, convention and declaration needs to be ratified by every member state. If a member state does not do so, it also is not bound by the agreement. In the case of the Conventions on World Heritage, it then also cannot take part in the nominations—and hence also not take part in the international glory that results from successfully being placed on the heritage lists.

The ICH Convention does not only result in online lists and printed representations published by UNESCO (cf. Hafstein 2009). As soon as they are officially ratified, nominations also create a new bureaucracy (this is of course also true for international or European agreements on trade, labor policy, human rights, and so forth). The heritage regime is reflected in measures that themselves can create further institutions, or in other words, in a bureaucratic apparatus. Such institutions might be established with reference to older protection or preservation bureaucracies, for in the end, cultural valuation mechanisms are not new. One need only think of movements protecting architectural heritage or other monuments, such as *Heimatschutz* (homeland protection) in the German-language areas of Europe. Such preservation-oriented values result, once institutionalized, in precise demands, for instance, in the preservation of nominated structures. Another example is the "musealization" of entire building complexes as it takes place in open-air museums since the late nineteenth century.[7] Each of these valuation mechanisms has a "rescuing intent" and each has brought forth a not small administering bureaucracy over time (Tschofen 2007). Bread, however, hardly needs rescuing—and neither did most of the other already successfully nominated foods. Again, one might thus wonder what propels actors to seek a nomination.

The world heritage regime unavoidably brings arguments by global actors into domestic discourses: cultural geographer Thomas Schmitt discussed this under the appropriate term (and title) *Cultural Governance* (2011, cf. Schmitt 2012), drawing on his research in North Africa. The centrality of these practices and scholars' theoretical focus away from considering them as a "meta" level activity become clear here. A specific heritage regime emerges, taking its cues also from

---

6   UNESCO's general conference admitted Palestine to be a member on October 31, 2011. The vote was understandably highly controversial, as Palestine is not a recognized, territorial state. On June 12, 2012, UNESCO conferred world heritage statues to the Church of Nativity and the Pilgrimage Route to Bethlehem, leading the USA and Israel withholding funding to UNESCO in 2012.

7   "Musealization" is a term coined by the German historian Wolfgang Zacharias (1990), subsequently taken on by German museum theorists such as Gottfried Korff or Martin Roth.

the type of state within which it unfolds.[8] This can be a very totalitarian and top-down process, as is currently the case, for example, in China or Cambodia. At the other end of the spectrum, there can be a federalist openness, as is currently evident in Switzerland.[9] The German bread registry, growing digitally since the spring of 2011, is also dealt with in a very open and participatory manner, but unlike in the Swiss case, the state has not decided as of yet to ratify the convention and the effort to convince politicians to do so is part of the rationale for the bread registry on the part of the major stakeholders, the German bakers' guild.

Germany, like various other European countries such as Ireland or Switzerland, has held back on this ratification for good geopolitical reasons. The ICH Convention came about not least because the list of monuments accepted into UNESCO's world cultural heritage has been hopelessly over-weighted by the global North, the (western) industrial nations. Precisely because countries like Germany, France or Italy can look back on a considerable history of monument preservation, it was easy for them to prepare monuments, building complexes and the like for the nomination process.

With the introduction of the ICH convention, the global South was to be given a chance to find international recognition for its cultural wealth. At the same time, potentially endangered traditions were to receive more recognition and through it also develop new dynamics. One can also remark in passing that behind this Convention stand decades of work to rethink and update the notions of culture and tradition used by UNESCO—as Valdimar Hafstein (2007), Richard Kurin (2004) have documented and analyzed. Thus UNESCO, too, has reached a less static, less closed definition of culture, though perhaps still not sufficiently so with respect to the idea of group cultures, whether in terms of nations or ethnic, indigenous or autochthonous groups. Yet it has also become clear to those responsible for world heritage that cultural practices change, and that nominations if not handled carefully, can lead to ossification. Not all actors in the heritage-making process work with a more flexible culture concept, however, and particularly at the local level, some stakeholders rely on antiquated theories of culture due in part to the time lag inherent to knowledge transfer, in part because some intellectually no longer tenable culture theories remain politically powerful and hence useful. Overall, matters have improved, but as any observer of international negotiations in the realm of cultural policy will witness, there are many definitions of culture and tradition, deployed depending on stakeholders' differential goals (Groth 2010).

The genteel restraint shown by European states with respect to the ICH came to end, not least with the spectacular nominations from the area nutrition and meals mentioned above. While Sarkozy's original announcement in February 2008 that French cuisine deserved world heritage status was greeted with a mixture

---

8   The Göttingen conference "The Heritage Regime and the State" (Bendix, Eggert and Peselmann 2013) examined precisely this confluence.

9   For a look at the ongoing process of compiling a possible ICH inventory in Switzerland, visit "Lebendige Traditionen" (2011).

of amusement and outrage by the international press, the food-nominations have created a certain degree of emulation and competition ("Sarkozy" 2008). It is not only the bakers in Germany whom one can regard as stakeholders in the effort to move Germany to join this UNESCO Convention. There are also other food-related custom bearers in Germany, such as the manufacturers of dumplings from Thuringia, supported by the dumpling museum, who have increased efforts to mobilize their parliamentary representatives (Vates 2011). Minister Aigner does not stand alone. The SPD parliamentary group, for example, published the following statement on June 30, 2011:

> It is finally time that Germany ratifies the UNESCO Convention on Immaterial Cultural Heritage. 136 nations have already ratified the Convention, including many of Germany's neighbors. With implementation of the Convention, Germany will be able to give special recognition to living traditions, practices, rituals and customs, as well as to the knowledge of traditional craft techniques. Even knowledge and practices in dealing with nature and the universe n belong to immaterial cultural heritage.[10]

But the mills of ratification grind slowly. Since late June of 2010, at least, representatives of various parliamentary groups have submitted motions to proceed more rapidly with the ratification.[11] Yet such motions need to be examined: What might the costs be? What institutional consequences could result? In other words, the discussion of ratification goes hand-in-hand with weighing and working out an appropriate regime for introducing this new procedure. Adopting the convention can, furthermore, not be done by the federal parliament alone: German law requires that the individual states be consulted and agree with any federal decision. It is thus not just national competition that plays a role, though the pressure to keep up with other nations is clearly evident at the grassroots level and exerts pressure on individual parliamentarians at the level of national politics. But in a time when crisis negotiations are being held to try to stave off the potential bankruptcy of Greece, Italy or the EU as a whole, the ratification of a UNESCO Convention—and the potential honoring of German bread—does not exactly have priority.

Regardless of what is playing out on the political stage, the *Zentralverband des deutschen Bäckerhandwerks* continues to add further varieties of bread to its online registry. Amin Werner, president of this modern day institution representing the once powerful early modern guilds, emphasizes the positive effect of this endeavor: to him, it is important to raise popular consciousness for the quality and craft inherent to artisanal breads. More and more industrial bread production takes the place of neighborhood bakeries, and if a bread registry

---

10   Published as SPD press statement NR, 794/2011, July 1, 2011.

11   E.g. on June 29, 2011, there was a proposal from representative of the bi-partisan governing coalition of CDU/CSU and FDP to ratify the convention, published as Drucksacke 17/Vorlagennummer 2415, AK6.

assists in raising the self-esteem of bakers, then perhaps a nomination as ICH might alert German bread consumers that the baking craft is a value worthy of artisanal rather than fast food prices (personal communication, Amin Werner, Nov. 28, 2011). Werner and the organization and profession he represents are thus ultimately concerned with the labor and food market—concerns that do not match the perhaps more regionalist or nationalist, not to say chauvinist, impulses of other ICH nominators. Before I turn to the truly burning issues for German bakers, I would like to take a brief detour to the scholarly antecedents of mapping cultural practices such as baking and the peculiar gap between scholarly effort and public use of such endeavors as evidenced in the bread registry.

Folk Atlases, Heritage Lists, Bread Registries, or: The Transience of Specialized Knowledge in Folklore Studies

Bread has been a topic of cultural research, though its omnipresence has also chastened researchers into frugality when it came to the topic. Camporesi's history of food's role in hungry people's actions and imaginations (1989) is just one of many examples of scholarly works taking bread as a metaphor for food's place in socio-cultural and political processes in general. There are a number of German bread studies (e.g. Eiselen 1995), some of them associated with open air museums which in turn are the kinds of institutions that naturally instrumentalize the scent and allure of bread as an inroad into regional pasts (Gentner 1991, Kaiser 1989). Bread museums—of which there are two major ones in Germany, one in Ulm and one in Ebergötzen—provide historical and cultural documentation as well, and there are the occasional dissertations focusing on bakers and their place in changing times, such as Koellreuter's study of bakers in the city of Basel since medieval times (2006).

But if one, as a cultural anthropologist or folklorist with European training, examines the bread registry map, she or he is likely reminded of the many large-scale atlases of folk culture produced in the inter- and postwar years, and will see certain similarities between the current efforts of the bread bakers' federation to create a bread registry and such scholarly inventorying efforts. The opening paragraphs of the German Wikipedia entry to the *Atlas der deutschen Volkskunde* reads as follows:

> Following the models of the surveys conducted by Wilhelm Mannhardt in 1865 and for compiling the *Deutscher Sprachatlas* [Atlas of German Speech], five-part questionnaires, sub-divided into 243 main questions, were sent out between 1930 and 1935 to 20,000 informants. Through a close connection between the ideologically charged categories of *Volk* and *Raum*, the idea was to establish that 'despite all variety and diversity (…), the German *Volk* was an indivisible unity'. (Fritz Boehm).

The results of the survey were published, until 1939 and in six installments, as 120 (uncommented) dispersion maps. Beginning in 1938, a further evaluation was conducted in the context of the *SS-Ahnenerbe* [SS-Ancestral Heritage]. After 1958, the survey answers were re-evaluated, and the maps in the new series received extensive critical commentary. ('Atlas' 2011)

The most expensive production supported by the German Research Foundation in the interwar era thus was searching to create a systematic inventory of regional folk culture. One should add that it was by no means the case that the Germans were the only ones to do so. It was just that in the political context of the Third Reich, the intent of territorial mapping of course served as legitimation for political aggression (Schmoll 2009). That inventories of this kind were returned to after 1945, and were even used as a means for international scientific cooperation indicates the "empty" potential of these information-gathering methods.[12] Their theoretical paradigm, which was worked out already in the last decades of the nineteenth century, searched evidence for hypotheses about origin and diffusion.

Using the approaches typical for the time, and wholly without digital techniques, Question 196 of this German atlas survey asked reliable local informants about bread-related topics. The goal was to systematically evaluate this information. But the collection resulted in far more information than there were, and are, researchers to evaluate it all. For it is not enough for a researcher to simply produce a geographical depiction: one needs commentaries for each map as well. In the case of bread, Günter Wiegelmann wrote the commentary to "daily bread," though he limited himself to bread spices and began with the sober assessment

The maps NF 11a-d show bread spices. This is drawn from the highly differentiated complex of questions asked which were related to daily bread. To include all aspects of Question 196 would have called for a long series of maps. This would have heavily freighted the entire cartographic process, one which must take many viewpoints into account both thematically and methodologically. Given the rich material gathered by the ADV survey, the entire complex surrounding bread could likely only be adequately addressed in a broadly conceived monograph. (1966–1982, 251)

Tough luck, therefore, for the bakers. The maps and studies necessary for submitting a cultural heritage nomination are simply not available from our discipline. Where the spices anise, caraway, fennel and dill were used in bread in the 1930s, and how far back one can trace their use, does not generate much by way of prose for a heritage application.

Nevertheless, it is precisely the nominations for world cultural heritage in the realm of food preparation and the culture of meals that one sees with

---

12 Friedemann Schmoll, University of Jena, is concluding a project on the European Atlas from 1920–1980; cf. also Rogan (2008).

particular clarity the connection between the documentation of culture that researchers carry out and the global valuation of culture itself. In its earliest incarnations, folkloric research was involved in the statistical survey of folk culture (Rassem 1951). This supported the territorial confines of the discipline, alongside the administering impulse the modernizing form the state had taken. A second, ideologically powerful pillar of folklore research, though, was the romantic-national glorification of agrarian and pastoral life. From this emerges a praising and thus valorizing if not upgrading, of cultural practices rooted in tradition. Territorial rootedness—as strange as it may sound—is also a basic principle of UNESCO. For while nominated monuments, landscapes and cultural practices belong to the heritage of mankind, their use and administration "belong" to the respective nation-state, as well as to those who in this nation-state emerge as the beneficiaries.

While cultural research has replaced veneration, as well as "saving by collecting," with various approaches marked by an—often critical—understanding of culture, the knowledge transfer of basic folklore terms and canonical knowledge to the public sphere has been quite successful. Voluntary associations as well as local museums have in many places taken it upon themselves to preserve customs of all kinds, and thereby provide a basic group of actors who not infrequently become involved in putting together UNESCO World Heritage dossiers. In the case of the planned bread-nomination in Germany, the bakers' guild has entrusted the Ulm bread museum with pulling together the requisite information (personal communication, Amin Werner, November 28, 2011). In the arena of cultural governance, what remains of the complex research techniques used to document culture is at times little more than registration. That in turn promotes lists, which fosters a not altogether productive competition, as Valdimar Hafstein has convincingly shown (2009). That the *Zentralverband des deutschen Bäckerhandwerkes* has adopted registering and mapping and is engaged in convincing the German parliament to adopt the ICH convention calls, however, for a further explanation, which I would like to turn to in the conclusion.

Why this is Not Only About Baking Buns: Craft and Immaterial Cultural Heritage

There are other systems available to add value to cultural products in the realm of food, particularly so in Europe. One needs to therefore ask why the German bakers opt for the heritage route and did not choose to use "Geographical Indication" (G.I.) as a means to help the many German types of bread receive greater recognition and thereby appreciate economically (Bramley and Kirsten 2007). A G.I. can be applied for with the European Union, and does not need to take the detour through state or national legislatures. Regional actors can apply directly to the EU, though they have to also reckon with being turned down. Varieties of sausage and cheese and even particular types of vegetables including the lamb's lettuce [*Feldsalat*] grown on the island of Reichenau, the asparagus from Schrobenhausen, or even—a baked good, after all!—gingerbread [*Lebkuchen*] from Nüremberg have all received such 'indications' related to geographic origin. While France has long had such a system (*appelation d'origine contrôlée)*, the "protected designation of origin" framework only came into effect in the EU in 1992. The justification for these instruments can be summarized as follows (cf. Bicskei et al. 2011):

- Agriculture can, in this manner, be diversified. For economically backwards rural areas, this can be a tool to help in economic development.
- Agricultural products receive a boost, as imitations are blocked or are at least marked as such.
- The measures are intended to protect consumers from fraud.

In addition to the advantage in the marketplace that one promises oneself through the emphasis placed on traditional and locally anchored forms of production, the corresponding manufacturing processes as well as the diversity of types (such as in varieties of cheese) is emphasized.

At the moment, there are three types of such protection, whereby the territorial-cultural localization is the most exclusive variant:

- Protected Designation of Origin: Here the quality of the product "is significantly or exclusively determined by the geographical environment, including natural and human factors." Production, processing and preparation take place in this specified geographical area and the product must be *entirely* manufactured in the designated region.
- Protected Geographical Indication: At least one level of the production must take place in the designated geographical area, so "the entire product must be traditionally and at least *partially* manufactured" there. There needs to be a connection between product and region, but that connection need not be exclusive.
- Protected Traditional Specialty: Here the product does not have to be produced in a particular or geographically limited region. However, it is important that its production "be manufactured using traditional ingredients or must be characteristic for its traditional composition, production process, or processing reflecting a traditional type of manufacturing or processing," and that it is "different from other similar products" (see "Protected Geographical Status" (http://en.wikipedia.org/wiki/Protected_Geographical_Status, accessed December 23, 2011).

This may be enough to demonstrate that these three categories are defined in ways that can lead to confusion. While manufacturers can, in fact, engage in delicate competition here, these differences are not nearly so easy for consumers to follow. It may need considerable public relations or translation (cf. Hegnes 2010) to convince manufacturers that this instrument brings a competitive advantage with it, and that consumers will actually pay attention to these labels, and through their purchases contribute to keeping certain manufacturing practices alive.

Yet bread is too large a category, and its manufacture is easy to move around. After all, within one year, the German Bread Registry accumulated more than 2,500 varieties of bread. Bread is not an exclusive product the way a smoked meat or a sheep or goat cheese might be. For them, consumers are willing to pay a little more. For bread, though, the opposite is true: one only need be reminded that it was an increase in the price of bread in 1789 that set off the French Revolution ....

For the bakers' craft, a Geographical Indication thus brings little or no economic gains. What is lacking, in fact, are not bread-buyers but bakers themselves. Beyond the pleasing smell of freshly-baked bread, apprenticeship and training to be a baker means hard physical labor and extremely early rising and working hours. There are, correspondingly, simply not enough apprentices who want to take up this craft, which furthermore cannot offer competitive wages.

The path to becoming a baker looks, ideally, like this:[13] The baker (according to the Handicrafts Code a "trade" with the obligation to become a "craftmaster") begins a three-year apprenticeship in a bakery enterprise headed by a master-baker. Nevertheless, he receives one to two days' instruction at a trade school to teach all that one does not learn working within the enterprise. At the end of three years, he must pass a test to become a journeyman, which is organized by the Handicrafts Chamber. The entrepreneur with whom the apprentice is training must be a member of that Chamber and pay dues to it. The Handicrafts Chambers (of which there are 53 in Germany) go back to, or are based on, local guilds. These guilds, however, are voluntary associations, mostly organized at the district level. This is what creates the conflicts between guilds and chambers. The guilds do what can be done on a voluntary basis; the chambers do what is mandatory and exercise sovereignty. But since the chambers lack the necessary specialized knowledge, they must repeatedly rely on the guilds. The guilds, however, are losing members, as more and more—in this case—bakeries are closing.

If the apprentice passes his test at the end of three years, he becomes a journeyman. After five years in a leading position, he may, as "senior journeyman," take over the running of an enterprise. Until now, the regular path required him to pass a test to become a master craftsman, one which included training in sales and personnel management. A journeyman should already have all the necessary specialized knowledge of his *métier* but the breakthrough in terms of business only comes with becoming a master craftsman. After that, he can train others and run his own firm. There are also special rules or exceptions for senior journeymen. Given the economic pressures, the practice at this point is to say "If the capacity is there, they should be allowed to run their own firm."

---

13 Thanks to Kilian Bizer and other colleagues at the Volkswirtschaftlichen Institut für Mittelstand und Handwerk (ifh) at the University of Göttingen for this depiction.

A 2011 study of craft and trades development found that "between 2000 and 2007 ... the number of those employed (and subject to social insurance provisions) in the bakery trades has fallen by around 25,000. The number of the marginally employed is increasing, the number of those with a training in the profession is decreasing" (Scholz 2011, 4). To this come the closure of smaller bakers and an increase of chain-stores. The president of the *Zentralverbandes des Bäckerhandwerkes* Werner estimates that about 300 bakeries annually go out of business. This, in turn, reduces the power and effectiveness of the traditional craft association. The erosion of guilds as well as guild associations in turn leads to a weaker position in wage negotiations—which translated means that their ability to convincingly argue for wage stability in the bakery trades has weakened (Scholz 2011, 1). The result: increased wage undercutting, precarious employment at low wages, and an absence of standardized wages.

From this perspective, it will hopefully have become clear that though people worldwide presumably love oven-fresh bread of all kinds of varieties, there is a real danger of losing daily access to this pleasure which has little to do with UNESCO but a great deal to do with the socio-economic valuation of craft and handiwork. Bakeries are not the only type of craft or trade that have an increasingly hard time finding apprentices. But there are few enterprises where the discrepancies between consumption and the conditions of production are so intrinsically obvious as in bakeries.

It should be evident why German bakers seek cultural valorization through a nomination as Intangible Cultural heritage. The baker's federation is using what is meanwhile a well-established strategy of increasing appreciation and esteem in an effort to ensure the future of its craft—not just in how it is carried out but also in how it is remunerated. UNESCO tries, using its Convention on ICH, to support endangered cultural practices. Compared to the cultural goods thus far acknowledged in the area of foodstuffs, the German bakers have every reason to draw attention to their craft and to the manifold products it brings forth. The German Mission in the United States featured the bakers' crusade in an online article in October 2011—surely a good indication that the German bakers have solid reasons to hope that the German parliament is well on the way to ratify the ICH convention ("German Missions" 2011).

Coda: In July 2013, Germany ratified the ICH convention. A first set of submissions on the state level will be discussed by a federal expert commission in the course of summer 2014 and an announcement as to whether the bakers succeeded to be recognized, in a first step, on the national register, is expected in December.

References

Bendix, R., A. Eggert and A. Peselmann (eds) (2013). *Heritage Regimes and the State*. Göttingen: Göttingen University Press.

Bicskei, M., K. Bizer, K.L. Sidali and A. Spiller. 2011. *Reformvorschläge zu den geografischen Herkunfsangaben der EU*. Unpublished Working Paper, Göttingen.

Bramley, C. and J.F. Kirsten. 2007. Exploring the Economic Rationale for Protecting Geographic Indications in Agriculture. *Agrekom*, 46(1): 69–93.

Broccolini, A. (2013). The Intangible Heritage Inventories in Italy: Between Community, National Politics and Anthropological Competencies. In *Heritage Regimes and the State*, edited by R. Bendix, A. Eggert and A. Peselmann. Göttingen: Göttingen University Press, pp. 283–90.

Camporesi, P. 1989. *Bread of Dreams*. Translated by David Gentilcore. Chicago: University of Chicago Press.

Eiselen, H. (ed.) 1995. *Brotkultur*. Köln: Dumont.

Gentner, C. 1991. *Pumpernickel—das schwarze Brot der Westfalen*. Detmold: Westfälisches Freilichtmuseum.

Groth, S. 2010. Perspektiven der Differenzierung: Multiple Ausdeutungen von traditionellem Wissen indigener Gemeinschaften in WIPO Verhandlungen, in *Die Konstituierung von Cultural Property: Forschungsperspektiven*, edited by R. Bendix, K. Bizer and S. Groth. Göttingen: Universitätsverlag Göttingen, pp. 177–95.

Hafstein, V. Tr. 2007. Claiming Culture: Intangible Heritage Inc., in *Prädikat Heritage. Wertschöpfung aus kulturellen Ressourcen*, edited by D. Hemme, M. Tauschek and R. Bendix. Münster: Lit Verlag, pp. 75–100.

Hafstein, V. Tr. 2009. Intangible heritage as a list. From masterpiece to representation, in *Intangible Heritage*, edited by L. Smith and N. Akagawa. London: Routledge, pp. 93–111.

Hegnes, A.W. 2010. Der Schutz der geographischen Nahrungsmittelherkunft in Norwegen als Übersetzungs- und Transformationsprozess, in *Essen in Europa. Kulturelle „Rückstände" in Nahrung und Körper*, edited by S. Bauer et al. Bielefeld: transcript, pp. 43–64.

Kaiser, H. 1989. *Das alltägliche Brot: über Schwarzbrot, Pumpernickel, Backhäuser und Grobbäcker. Ein geschichtlicher Abriss*. Cloppenburg: Museumsdorf Cloppenburg.

Kirshenblatt-Gimblett, B. 1995. Theorizing Heritage. *Ethnomusicology*, Vol. 39 (1995). Illinois: University of Illinois Press, 367–80.

Kirshenblatt-Gimblett, B. 2004. Intangible Heritage as Metacultural Production. *Museum International*, 56 (1–2): 52–65.

Koellreuter, I. 2006. *Brot und Stadt: Bäckerhandwerk und Brotkonsum in Basel vom Mittelalter bis zur Gegenwart*. Basel: Schwabe.

Kurin, R. 2004. Safeguarding Intangible Cultural Heritage in the 2003 UNESCO Convention: A critical appraisal. *Museum International* 56(1–2): 66–77.

Rassem, M. 1951. *Die Volkstumswissenschaft und der Etatismus*. Graz: Akad. Druck- und Verlagsanstalt.

Rogan, B. 2008. The Troubled Past of European Ethnology. SIEF and International Cooperation from Prague to Derry. *Ethnologia Europaea*, 38(1): 66–78.

Schmitt, T.M. 2011. *Cultural Governance. Zur Kulturgeographie des UNESCO-Welterberegimes*. Heidelberg: Franz Steiner Verlag.

Schmitt, T.M. 2012. *The Heritage Regime*. London: Routledge.

Schmoll, F. 2009. *Die Vermessung der Kultur: Der „Atlas der deutschen Volkskunde" und die Deutsche Forschungsgemeinschaft 1928–1980*. Stuttgart: Steiner.

Scholz, J. 2011. *Strukturveränderungen, Tarifsituation, Einkommensentwicklung und gewerkschaftliche Revitalisierungsansätze im Handwerk*. Unpublished IHF Report: July 28.

Siebeck, W. 2011. *Ofenwarme Träume*. [Online]. Available at: http://www.zeit. de/2011/16/Siebeck (Accessed 22 November 2011)

Simon, M. 2003. "Volksmedizin" im frühen 20. Jahrhundert. Zum Quellenwert des Atlas der deutschen Volkskunde. *Studien zur Volkskultur in Rheinland-Pfalz*, Band 28. Mainz am Rhein: Gesellschaft für Volkskunde in Rheinland-Pfalz e.V.

Tauschek, M. 2009. Cultural Property as Strategy. The Carnival of Binche, the Creation of Cultural Heritage and Cultural Property. *Ethnologia Europaea*, 39(2), 67–80.

Tornatore, J.-L. (2013). "Anthropology's Payback: 'The Gastronomic Meal of the French.' The Ethnographic Elements of a Heritage Distinction." in *Heritage Regimes and the State*, edited by R. Bendix. A. Eggert and A. Peselmann. Göttingen: Göttingen University Press, pp. 341–65.

Tschofen, B. 2007. Antreten, ablehnen, verwalten? Was der Heritage Boom den Kulturwissenschaften aufträgt, in *Prädikat Heritage. Wertschöpfung aus kulturellen Ressourcen*, edited by D. Hemme, M. Tauschek and R. Bendix. Münster: Lit Verlag, 19–32.

Vates, Daniela. 2011. "UNESCO—Klösse als Weltkulturerbe." *Frankfurter Rundschau*. November 10.

Wiegelmann, G. 1965. Das tägliche Brot: Brotgewürze, in *Atlas der deutschen Volkaskunde. Neue Folge. Karten NF 44a–44d. Erläuterungen Band II*, edited by M. Zender. Marburg: Elwert, pp. 251–75.

Wiegelmann, G. 1995. Täglich Brot, in *Brotkultur*, edited by H. Eiselen. Köln: Dumont, pp. 231–43.

Zacharias, W. (ed.) 1990. *Zeitphänomen Musealisierung. Das Verschwinden der Gegenwart und die Konstruktion der Erinnerung*. Essen: Klartext-Verlagsgesellschaft.

Online References

"Atlas der deutschen Volkskunde" http://de.wikipedia.org/wiki/Atlas_der_deutschen_ Volkskunde (accessed November 28, 2011).

"German Missions in the United States—German Bread" http://www.germany.info/ Vertretung/usa/en/__pr/GIC/2011/10/12__Bread__PR.html, (accessed November 28, 2011)

"Korea.net" http://www.korea.net/news.do?mode=detail&guid=56565 (accessed December 20, 2011)

"Lebendige Traditionen" http://www.lebendige-traditionen.ch/index.php?action= screen&size=1024, (accessed December 20, 2011).

"Panel wants UNESCO recognition for Japanese food" http://ajw.asahi.com/article/ cool_japan/culture/AJ2011111117101 (accessed December 20, 2011)

"Sarkozy wants UNESCO to honor French cuisine," http://news.oneindia. in/2008/02/24/sarkozy-unesco-honour-french-cuisine-1203851100.html, (checked December 26, 2011).

"Vorschlag von Bundeslandwirtschaftsministerin Aigner" http://www.rp-online. de/politik/deutschland/aigner-deutsches-brot-als-weltkulturerbe-1.2289408 (accessed on November 28, 2011)

"World Public Health Nutrition Organization" http://www.wphna.org/2011_mar_ hp1_unesco.htm (accessed Dec. 20, 2011).

Chapter 13

The Mexican and Transnational Lives of Corn: Technological, Political, Edible Object

Erick Castellanos and Sara Bergstresser

We all came from corn.
We have our roots in the earth,
We all reach for the sky,
And just like corn,
We are all little kernels of different colors.

Mam Maya man, Mahwah, NJ

Spending time in rural areas of Mexico, one comes to appreciate the importance of corn. The planting is carefully planned, the fields are diligently tended, and the harvests are joyously celebrated. Often tortillas are still made by hand; women pat the dough into flat circular discs, which they cook on a pan over an open fire. These appetizing morsels vary in color depending on the place and time of year: white varieties that grow at high altitudes or in warm humid climates, the indigo corn of northern Mexico, or the crimson maize of central Mexico. Men retell stories about the central role that corn plays in human existence. It is difficult to believe that their staple is the very same mechanically harvested grain that towers below grain elevators in Iowa, destined either to feed thousands of cattle or to be factory-processed into sugar or fuel. In contrast to the diversity found in Mexico, all of the Iowa corn is genetically identical; it was scientifically designed to maximize production. Corn's versatility has made it one of the most important staples in the world, and because of its ubiquity, people have come to invest it with a diverse set of meanings.

A scientist in Ithaca, New York, a northern Italian grandmother, a Wall Street commodities trader, and a Mexican *campesino* will all have different answers to the question "What is corn?" For each of them, corn is much more than a simple food product; the answer depends on the broad social, scientific, economic, political, and historical contexts in which a person lives. Since corn plays such a central role in individual lives, community traditions, national politics, and international relations, its polysemic nature should not come as a surprise.

To understand the multiple perspectives and symbolic meanings of corn, it is useful to analyze the metaphors that surround it. Eivind Jacobsen identifies three dominant tropes that actors employ when engaging in food rhetoric: food as nature, food as commodity, and food as culture. She explains, "References to various tropes and connotative fields imply different conceptions of time and relevant spaces. This opens up the possibilities for different ethical or moral considerations, as it makes different consequential universes relevant" (2004: 62). She goes on to point out that "actors tend to purify distinct aspects of food (as signifier), claiming a specific trait to be its essence, and hence ignoring other aspects or taking them for granted" (63).

The food-as-nature trope focuses on the sources of food and on the nutrition it provides to humans. Food can serve as a connection to nature or as way to overcome its limits. Within this trope, the role of science in the production of food varies: it can be a mechanism of progress that enhances what nature provides, it can be a corrupting instrument that taints the purity of nature, or it can be an apparatus of mediation to draw on nature to improve the human condition.

The global dominance of capitalism is intertwined with the trope of food-as-commodity. From this perspective, food is a neoliberal good, which is to be produced, distributed, and sold through markets. Industrialized nations and international organizations often coerce low-income countries into accepting food as a globally exchanged commodity through economic incentives, political negotiations, and ultimately by force. The vision of food as commodity also tends to promote standardization, for food produced in one place should not be any different from food from another. However, many people have begun to object to and resist the move towards indistinguishable food.

When food is seen as culture, it becomes a tangible (and perhaps a visual, olfactory, audible, and gustatory) symbol of a specific group of people. This view is predicated on the belief that food can encapsulate a group's history, traditions, values, and beliefs. Thus food becomes the avenue through which those who possess power or authority demarcate social and cultural boundaries, both in temporal and geographic space (see Bourdieu 1987, Douglas 1966, Elias 1978). The trope of food-as-culture often links back to the trope of food-as-nature, and culture becomes portrayed as primordial and natural. For example, our research in Italy revealed that residents often saw their culturally distinct food as being rooted in the soil of Bergamo (see Castellanos and Bergstresser 2006). Building on this concept, we would suggest that there is another trope that draws on link between culture and geography: food as heritage. From this perspective, food becomes both a symbolic and economic resource. It is a way to link the past with the present and create a connection between the present the future. Envisioning food as heritage becomes a way to authenticate and validate the existence and the practices of a culture and becomes a way to commodify the culture itself (see Lindholm 2008).

The metaphors used to understand the role of corn in people's lives can be understood through these four food tropes. In this chapter, we will trace the journey corn has taken from central Mexico across the globe and how its perceptions have fit into the rhetoric of nature, commodity, culture, and heritage. We will focus on the conflicts and contrasts between these various rhetorical approaches, focusing on the social and economic roles of corn in Mexico. Last, we will illustrate how all four rhetorical tropes have converged as Mexico debates the future of corn as a genetically modified organism (GMO) in the country.

Corn in Mesoamerica

> Without corn, we are nothing. (Tzotzil Maya woman, Chiapas)

Ethnobotanists, archaeologists, and geneticists have traced the origins of corn to Mesoamerica; some believe that it was first domesticated in the highlands of central Mexico (Fussell 1992), while others believe that its roots are in the Balsas river valley of eastern Mexico (Doebley 2004). Through fossilized pollen samples, scientists have determined that corn was domesticated from teosinte, a variant of grass, about 9,000 years ago. By the time Europeans arrived in 1492, corn had become a ubiquitous staple throughout the Americas. The word maize comes from the term Columbus heard the Arawak island's inhabitants use to refer to it: *mahiz* (Fussell 1992).

The value of corn lies in its versatility as a grain. It can function as an immediate food source when eaten freshly picked, and it can also become a commodity when dried, because in this form it can be easily transported and stored. Maize can also be mashed and fermented, and this substance can then be brewed or distilled. The husks can be woven into rugs or twine, the leaves and stalks are excellent fodder for domesticated animals, and the cobs can be burned for fuel. Beyond its multifunctional practicality, corn's genetic versatility and nutritional offerings have cemented its place in human history.

As valuable as corn is to humans, the grain has also pegged its own future survival on humans. While there are grasses that resemble maize, corn itself does not exist in the wild, and it cannot reproduce without human intervention; yet it is extremely adaptable to different climates and soils, it grows quickly, and offers high yields. Corn, thus, makes up for the care and reproductive assistance it receives from humans by out-performing other grains and staples, by adapting to the climate wherever people may take it, and by addressing the multiple needs of its human custodians.

The ongoing importance of maize in the lives of Mesoamerican indigenous people is evident in the centrality of corn in myths, iconography, and spiritual beliefs. Archaeologists frequently find representations of corn in areas that were used for ceremonial and religious purposes. These images reveal that to them corn was both the giver of life and the connection to the supernatural. The Maya of southern Mexico and Guatemala believed (and continue to believe) that humans came from corn. Their creation myth, found in the *Popol Vuh*, details the gods' unsuccessful attempts to fashion humans out of animals, mud, and wood. Subsequently, the gods molded humans out of corn meal dough (Tedlock 1996). Maya we have spoken with in Yucatán, Chiapas, and Guatemala continue to retell this story and refer to it when describing their relationship to and affinity for corn. Moreover, contemporary indigenous groups still respect and honor the crop, often in ceremonies that draw from both Christian and indigenous traditions. We have observed the use of corn at ceremonies held at ancient Mayan religious sites, Catholic churches, and syncretic temples that draw from both belief systems, such as the former Catholic Church in San Juan Chamula in Chiapas. Our discussions with indigenous peoples throughout Mexico reveal how they see corn as the foundation of their culture and heritage.

Wheat vs. Corn

When the Spanish colonized the Americas, they set out on the massive mission to impose their political, economic, religious, and cultural way of life on the indigenous peoples. Among these changes was the attempt to impose the consumption of European food staples, in particular, wheat. Spanish clerics believed that wheat was the avenue through which they could replace indigenous corn deities with the Christian God (Pilcher 2006). Beyond being the foundation for the European diet, wheat (and the bread made from it) was symbolically important, especially in religion. During the Eucharist, the communion wafer (unleavened wheat bread) undergoes transubstantiation when it is consumed: that is, it becomes the body of Christ. Canon Law specifies that only wheat may be used for communion. Even today, the Catholic Church will not allow substitutions of wafers made from other grains for people who have wheat allergies or Celiac disease (see Crane 2003). Thus, wheat was not only the grain of civilization, but it was also the medium through which one could commune with God.

The attempt to substitute wheat for corn during the colonial period failed because the indigenous populations were unwilling to consume wheat and Mexico's climate was not suited for its growth. The result was that two parallel food systems emerged in Mexico. These systems followed the divided social structure of the time, which was determined by class and race. Detailed racial categories and hierarchies were developed during the colonial era, where peninsular-born Spanish were at the top, followed by full-blooded Spanish who were born in the colony (*criollos*), then by those with mixed Spanish and indigenous blood (*mestizos*), and indigenous (and African) peoples were located at the bottom.[1] Within the racialized hierarchy of power, those at the top consumed a wheat-based diet, while those at the bottom continued to eat meals based on corn. In other words, the Europeans and *criollos* ate bread, while the *mestizos* and indigenous people ate *tortillas*. The Spanish began to attribute European superiority and civility to their consumption of wheat and "Indian" backwardness, feebleness, and susceptibility to illness to their corn-based diet. These perceptions, of course, ignored the brutal conditions most indigenous people were forced to live and work in and the catastrophic impact the various pathogens the Europeans introduced into the Americas.

---

1   The system was actually much more complex because there were categories for subtle gradations of heritage, such as the child of a *criollo* father and a *mestiza* mother. There are detailed *casta* charts that illustrate the matrix of racial combinations, indicating the relative social position of each category.

This colonial dietary pattern has continued until present day, where wealthy Mexicans still tend to consider high cuisine as coming from Europe and not including corn (or hot peppers) within its ingredients. This reflects the historical tendency of wealthy and the middle classes to see their culture and heritage as coming from Europe and minimizing their ties to contemporary indigenous peoples. Throughout our travels we have observed the abundance of restaurants catering to the middle class that offer bland food with a breadbasket in the middle of the table. Dishes such as spaghetti, caesar salad, and club sandwiches appear on the menu, whereas typical Mexican dishes are not listed or perhaps are hidden on the last page. It must be noted, however, that over the past 20 years or so there has been a "rediscovery" of "traditional" Mexican cuisine—an issue we will address later.

Corn in Modern Mexico

While corn was spreading throughout the world, it remained a key part of life in Mexico, especially among *mestizos* and indigenous populations. The production of corn as a food source had a great impact in social and gender relations. Relying on maize as your primary source of food is very work-intensive for both men and women. However, tasks were divided up strictly along gender roles. Men worked the fields planting, growing, and harvesting the corn, while women worked the kitchen cooking, grinding, and serving corn-based food. These tasks shaped gender identities and roles throughout rural and indigenous Mexico.

For women, the conversion of corn into a *tortilla*, the Mexican staple found at almost every meal, was a process that could last up to five hours (Pilcher 2002). A woman usually started the night before cooking the corn in a mineral lime solution to make *nixtamal*.[2] She then woke up several hours before sunrise to grind the corn on a *metate*, a grinding stone made of volcanic rock that is rectangular, slightly concave, and stands on three short legs. Once the corn was ground, she added water to it to form *masa*, or corn dough. The dough was shaped into patties by hand and heated on a *comal*, a flat earthenware or iron pan, either before or during the meal. The process had to be repeated every day because *tortillas* turn hard soon after they are heated and become unpalatable. The *masa* cannot be saved either, for it begins to ferment after about a day.

Through this hard work, women not only fed their families, but they also constructed and maintained their identities as wife, mother, and caretaker. They also attained status within their families and in the community. Mexicans still joke

---

2 Soaking the corn in the lime solution releases both niacin and key amino acid proteins that along with the proteins found in beans and squash can allow a population to eat a mostly vegetarian diet without sacrificing nutritional balance. Populations across the world began to suffer from *pellagra*, a B3 (or niacin) deficiency when they adopted corn as a staple because they skipped this step in the preparation of their corn (Warman 2003).

Figure 13.1   Tortillas being made on a *comal* over a fire. Chiapas, Mexico

Source: Photo by Erick Castellanos.

that a woman cannot get married if she does not know how to make *tortillas* from scratch. When technology that would ease the process began to spread throughout Mexico, it had profound social effects and it transformed gender roles.

The first technological change came in the form of mechanical mills in the 1920s and 1930s where women could take their corn to be turned into *masa*, saving them hours of grinding on the *metate*. However, many women were hesitant to make use of the mills because of concerns about the cost and how it might change their roles within the family. Men also believed that without working on the *metate*, women would become lazy and promiscuous (Pilcher 2002). Nevertheless, more and more women over time began to make use of the mills.

The next major technological advance was the development of fully integrated *tortilla* machines that could grind the *nixtamal*, press out the *tortillas* out of the *masa*, and cook them on a moving conveyor. These *tortillerías* began to appear in the late 1940s, and by the 1970s, the machines were found throughout the country. Unlike a solitary woman working with a *comal*, these machines could produce several thousand *tortillas* an hour (Pilcher 2002). Some Mexicans still prefer the taste (and texture) of homemade *tortillas* over those that were mechanically produced. This was evident during a recent trip to small towns in

Figure 13.2   A *nixtamal* mill where women take their corn to be ground. Cajolá, Guatemala

Source: Photo by Erick Castellanos.

Chiapas and Guatemala, where we observed shops where the mechanical mills had a rudimentary *Molino Nixtamal* painted by hand over the main entrance. The presence of these shops indicates that women in those towns still take their corn to be milled and take the *masa* home to make their *tortillas* by hand, probably to the satisfaction of their husbands. In bigger towns, however, finding a mechanical mill is rare.

As result of the proliferation of technology, over a span of 20 years, most Mexicans went from eating handmade and homemade *tortillas* to those made at a *tortillería*. Freed up from the labors of making *tortillas* from scratch, women now had time to dedicate to other activities. Consequently, there was a major shift in the roles of women, in both domestic and public spheres. Some began to work in the fields with men, while others engaged in commercial activities to supplement the household income, and yet others ventured into the workforce.

Changes in *tortilla*-making did not end with *tortillerías*. In the 1950s, the Mexican government addressed the country's growing food needs by refining corn on a larger scale. The goal would be to create *masa harina*, or corn flour, which could be made into *tortillas* simply by adding water. Scientists and government officials also believed that *masa harina* could be enriched with vitamins, thus

**Figure 13.3   A typical *tortillería* that produces tortillas made from 100%
                MASECA. San Cristobal de las Casas, Chiapas, Mexico**
Source: Photo by Erick Castellanos.

ensuring better nutrition for the population (this was never done, however).
Two entities, one public (Maíz Industrialisado, S.A.—MINSA) and one private
(Molinos Azteca, S.A—MASECA), collaborated to find the formula for this flour,
which they achieved by the late 1950s. Over the next 40 years, *tortillerías* began
to replace corn with *masa harina*, and the supply was controlled by MASECA.[3]
The change in tortilla manufacture is visible within the landscape of most Mexican
towns, where the *tortillería* was once a nondescript shop on a busy street or even
located in the main square. Today they are typically painted white with the green
and white logo of MASECA covering a good portion of the exterior wall.

While many Mexicans lament the loss of better tasting homemade *tortillas*,
they still obtain 40 percent of their calories from them (Pollan 2006), illustrating

---

3   In the 1990s the Mexican government sold Minsa to rival Maseca (Pilcher 2002).
More recently Maseca has changed its name to Gruma and has become a transnational
corporation. Gruma is the largest producer of corn flour and *tortillas* in the world. They
sell 30 varieties of corn flour and have subsidiaries in China, England, Venezuela, Central
America, and the United States (Hoover 2012).

the central role that corn continues to have in their daily lives. The symbolic value of the tortilla has proven particularly resistant to manipulation, even while the substance of the tortilla itself has undergone multiple waves of manipulation, both physical and ideological. Though this staple food may now be made by machines rather than female hands, it persists as a central image in the conceptualization and depiction of Mexican heritage.

Corn, the "Green Revolution," and Neoliberalism

In the decades following the end of the Mexican Revolution in 1924, the ruling party undertook the mission of institutionalizing the goals of the revolution through nationalist projects: appropriation of key industries from foreign investors, redistribution of land, and modernization of the country. Many of these projects were framed through a "backward projection" narrative that rooted modern Mexico's identity in the ancient empires of Mesoamerica (Bartra 2002). Part of the national story was that Mexico was a "maize civilization" and the *campesinos* were the stewards of this heritage (Richard 2008). Nationalist rhetoric thus linked the past to the present and the future through corn and its growers. Despite the desire to advance post-revolutionary Mexico through self-reliance, Mexican leaders soon faced the reality that it needed foreign assistance to address its economic and social needs. Mexico needed to develop ways to feed its quickly growing population.

In 1943, Mexican officials began a joint research program with the U.S. that aimed to solve Mexico's food problems through higher agricultural productivity (Fussell 1992). At the time, the U.S. was riding a wave of progress in corn production. Americans discovered the productivity of corn and its potential as an economic enterprise because of its adaptability and variability. Of particular interest was how certain hybridized types were highly productive in the first generation, but virtually useless in subsequent generations. This meant that farmers would constantly need to purchase new seeds, which was of great value for agricultural business, but not for farmers. Another benefit of growing genetically identical corn was that it lent itself to mechanization on the farm, which allowed for even greater productivity.

American scientists saw change as improvement and progress, and they were naturally drawn to corn because its variability made it open to modification; to them progress meant an increase in the efficiency of production, which they achieved. However, the increase in efficiency came at a cost. One was the destruction or loss of thousands of varieties of corn as a result of hybridization and standardization of production; the other was the eventual over-production of corn. In addition to the industrialization and modernization of the food supply another force had begun to penetrate the lives of Mexican and their relationships with corn: neo-liberalism.

Facing possible shortages of food, Mexico looked to U.S. science and technology to try to address the problem. The Mexican government, at the encouragement of the U.S. government and American companies, began to promote the use of technology and fertilizers in the farming of corn. Most of these technologies and chemical fertilizers were then imported from the U.S. The Mexican government had a particular interest in increasing productivity and lowering costs in the farming of corn because it provided food to poor communities. Food assistance had become a way for the government to keep the rapidly growing urban populations under control and to retain the political support of rural communities.

Policies that promoted the use of American technology hurt small farmers. They could not afford to purchase chemical fertilizers and tractors, nor could they install the irrigation requirements that some of the new hybrids required. Moreover, the Mexican state food agency that was created to help small farmers compete in the marketplace switched its focus to developing larger commercial growers (Pilcher 2002). More and more government corn purchases for its food programs for the poor came from the large growers rather than from the small farmers. As overall production increased throughout the country, the price of corn dropped, which forced small farmers out of the market. As small farmers from all parts of the country stopped growing corn, many varieties began to be lost. These farmers knew which varieties were appropriate for their respective climates and soils. When they left farming, those local varieties were no longer planted and thus lost.

The "Green Revolution" of increasing productivity through the use of science and technology worked. By 1980, Mexico had become food self-sufficient even though its population had grown tremendously over the course of the twentieth century (Weis 2007). However, Mexico was also deep in debt and at risk of defaulting on its financial obligations. International organizations, such as the IMF and the World Bank, and the U.S. government offered assistance, but only under the condition that Mexico undertake structural adjustment policies. These policies required a decrease in government spending and the opening of Mexican markets to foreign investment and products.

Throughout the 1980s and the early 1990s, the Mexican government cut food subsidies, stopped imposing price controls for *tortillas*, ended redistributive land reform, and promoted consolidation in agriculture (see Nash 2002, Pilcher 2002, Weis 2007). Mexican farmers were encouraged to switch production from corn to fruits and vegetables for export. The goal was to farm what was profitable, not what was necessary to feed the population. Corn and other grains would then be imported from countries that could produce them at a lower cost. This was a boon for the U.S., which is always looking for markets for its excess corn. The corn eventually would flow south without impediment thanks to the dismantling of trade barriers through NAFTA in 1994.

The increased corn imports from the U.S. had a vast impact on Mexico. In the late 1990s, maize made up more than half of Mexico's crops, supported three million households, and accounted for two thirds of the country's caloric intake. More than 60 percent of corn growers farmed five hectares or less, usually without irrigation. Moreover, because scientific hybridization had not yet developed high-yield varieties for the diverse and multi-crop environments found throughout Mexico, about 80 percent of the land was planted with native varieties (McAfee 2008).

In the 10 years after NAFTA, prices paid to Mexican corn farmers plunged more than 50 percent, Mexico's corn imports from the U.S. tripled, and it is estimated that one million people were squeezed out of agriculture (Nadal and Wise 2004, Weis 2007). Maize growers began to refer to the countryside as "disaster areas" because of economic crisis that ensued. In order to differentiate the corn disaster from the natural disasters that strike Mexico and to highlight the role the local and national leadership had in precipitating them, farmers began to use the term "government disaster" (Richard 2008). The displaced former farmers were forced to migrate either to large urban centers or across the border to the U.S.[4] Those who remained in the rural areas, however, continued to grow corn despite the fact that the cost of farming maize was higher than the price they could ever get for it. Overall production and total area planted remained unchanged in Mexico (McAfee 2008). Defying economists' expectations, Mexican farmers engaged in "irrational behavior" by not decreasing their production when prices decreased. Small farmers were willing to subsidize their own corn, either by drawing on profits from other crops or by relying on remittances from relatives in cities or abroad, because they gained a sense of security by doing so. Moreover, many of these farmers take great pride in their locally adapted and diverse varieties of corn, and they do not wish to give up on them. As Fitting states, "Farmers grow corn for a multiplicity of reasons, some formulated and articulated, others not … They may also grow corn because it is a tradition or a taken-for-granted part of rural life" (2011: 107–8). In sum, heritage was more important than economics.

Corn and GMOs

Despite the displacement of farmers, the rising cost of corn products, and the dissatisfaction with the new tasteless industrial *tortillas*, what catalyzed the country to protest the new neoliberal regime was the discovery of transgenes in Mexican varieties of corn in 2001. Mexicans from all segments of society began to debate whether GMOs posed threat to corn, to the cultural patrimony of the country, and to biodiversity. At the time, it was not legal to grow GM varieties in Mexico, but they could be legally imported. Unlike GMO debates in other countries, where

---

4 Many American politicians who campaign against undocumented immigration from Mexico often fail to recognize that it has been U.S. foreign and trade policy that is partially responsible for the increased migration from Mexico.

the focus was on all transgene production, what was at issue in Mexico was the purity of corn itself. There were already transgene varieties of soybean and cotton in Mexico since 1997, and these crops did not elicit the same widespread outcry.

The debates regarding GMOs revolved around issues of the environment, public health, modernization, productivity, global competition, culture, heritage, rural livelihoods, and questioning the authority of science in determining the future of corn in Mexico. In response to a public request by Mexican civil society organizations and to the broader public uproar, the Commission for Environmental Cooperation of North America (CEC), a body within NAFTA created to address environmental concerns, began an investigation of the issue and released a study: *Maize and Biodiversity: The Effects of Transgenic Maize in Mexico* (Commission for Environmental Cooperation 2004). The report not only considered the potential risks and benefits of GMO corn, including the effects on the environment and health, but it also included discussion of social and cultural effects GMO may have on the livelihoods and daily lives of those who rely on corn. Its recommendations included reducing corn imports into Mexico, labeling GMO corn, creating a system for the conservation of local varieties, and forming a monitoring system to determine levels of contamination. The U.S. criticized the report and rejected its recommendations. However, the broader scope of the study, which moved beyond scientific and health issues to address cultural and social concerns, was important in recognizing the broader meanings and impacts of corn.

Following these debates, Mexico passed legislation in 2005 that addressed biosafety and GMOs. The law recognized the need to protect plants that originate from Mexico, including corn. However, the law did not establish a mechanism through which these protections could be enforced, and anti-GMO activists soon began to refer to it as "Monsanto's Law," because they believed the law would do little to slow the giant agribusiness' push to flood the Mexico with GMO varieties of corn.

One of the groups that led the protests against transgene maize and the criticisms against "Monsanto's Law" was *In Defense of Maize*, a coalition of *campesino*, environmental, and indigenous rights activists that formed in 2002. The group put forth several demands through a manifesto published in 2006. First, maize needed to be protected as an issue of national security. Second, they demanded that any policies regarding GMOs needed to consider multiple perspectives and opinions, including those of rural and indigenous communities, before being approved. Third, they wanted funding for an independent monitoring agency. Finally, they called for a moratorium on the cultivation of GMO corn.

On the other side of the debate were corn producers from the northern parts of Mexico, who tended to have large-scale production farms. They supported the use of GMOs because they claimed that Mexico had one of the lowest maize yields in the world. They asserted that prohibiting GMOs would condemn the country to backwardness and marginalization (Fitting 2011). These producers continue to have the backing of U.S. officials who point out the promise of transgenic crops in addressing issues of underproduction, poverty, and hunger. Both producers and

U.S. officials tend to ignore the fact that GMO corn has had mediocre results in increasing production in a cost-effective way in the U.S. (McAfee 2008). The U.S. promotion of GMO crops stems from the strong influence of agribusiness on American trade policy, including the interests of the producers of transgenetic seeds. About six multinationals, with Monsanto leading the pack, dominate the markets in farm inputs, including seeds, fertilizers, herbicides, and pesticides (McAfee 2008). The promotion of GMOs abroad by the U.S. is part of a larger project that advocates the liberalization of food trade. Free trade advocates contend that the countries of the global south will benefit by importing food from the more efficient U.S. agricultural system, which has already harnessed science to increase productivity.

Biofuel from Corn, Price Destabilization, and the *Tortillazo*

Rather than improving their lives and lowering the cost of food, Mexicans in the first decade of the twenty-first century discovered that depending on imported corn exposed them to market-based commodity price fluctuations. Higher global demand as well as the new uses of corn for the production of biofuels led to a spike in the price of corn. The U.S. Energy Policy Act of 2005 mandated that all gasoline sold in the United States would contain 7.5 billion gallons of renewable fuels by 2012, and incentives were put in place to help that goal come about. From 2005 to 2006, ethanol production in the U.S. increased by one billion gallons, for a total of five billion gallons (Westcott 2007). Increasing demand for corn from the fuel industry diverts corn from other markets, including export; countries newly dependent on U.S. corn for food were increasingly competing with U.S. consumers' growing appetite for ethanol.

At the same time, Mexico began to follow the dictates of the IMF, and government subsidies and price controls disappeared. Once a price-controlled staple food, *tortillas* became subject to the fluctuations of the market, and prices began to rise. Many working-class Mexicans found that they could no longer afford their staple food. In January 2007, the *Tortillazo* (literally, the *Tortilla Coup*) hit Mexico. In the span of a month *tortilla* prices rose 70 percent in Mexico City and up to 150 percent in some parts of the country (Richard 2008, Suárez Carrera 2010). The sudden increase led to an unprecedented food crisis and people took to the streets. Protesters cried, "We are the people of the corn, and now we import it from the North!" (Weis 2007: 111). Civic organizations came together under the banner, *Sin maíz no hay país!* (Without Corn there is no Country!). The issue was front-page news across the country and the government was forced to intervene. In April, President Felipe Calderón signed the "Act to Stabilize the Price of Tortillas." While this action stopped the increase in prices, critics point out that it maintained the price at the new levels, which was a hardship given that wages did not increase. Moreover, part of the legislation created benefits for some of the large food companies in Mexico. In particular, Bimbo, the country's largest bread producer, was given incentives to produce more flour *tortillas* in an attempt to get more people to switch from corn, reducing the demand and eventually the price pressures of corn *tortillas* (Suárez Carrera 2010).[5]

The experience of the *Tortillazo* illustrates that Mexico is dependent on corn and that market pressures rather than changes in consumption behaviors have lead to social and civic action. It also suggests that Mexican officials need to consider more than economic concerns when developing future food and agricultural policy. In particular, the role of corn as a symbol of and vehicle for heritage turned out to be more important than the market-minded had anticipated. Far from being content with the rules of the market, Mexicans prioritized the meanings of the tortilla over the monetary gains promised by advocates of the neo-liberal worldview.

Corn, Food Sovereignty, and the Re-Invention of Tradition as Gourmet

Some say the corn is blind ...
Perhaps, the little corn may be blind, but it has life.
The bean has a little eye and it can see; so it can look out for itself.
Corn, however, we need to care for ... (Mixtec Man, Oaxaca)

Over the past 50 years, changing economic and political contexts have lead to a change in Mexican food policy. Following the Second World War, Mexico's major concern was food security: having enough food available to all of its citizens. Moving towards the end of the twentieth and into the twenty-first century, the government's perspective on food has evolved to one of food as a resource and commodity. Initially the government's concern for increasing productivity was based on the need to feed a growing population. Over time, however, what was at stake became an issue of competitiveness on the global market, and ultimately profitability. During this transition, the connection between food and people was replaced with the link between food and income. Over time, however, those affected by the changes began to question the risks taken and sacrifices made for the sake of profit, especially when that profit would only benefit a select few who did not share in the burdens of change. It is in this context that farmers, consumers, and environmentalists see the growing threats to corn. They view the scientific modifications to corn itself and the economic transformations of how it is grown not only as a threat to their way of life, the environment, and to the food they consume, but also as a threat to the cultural patrimony of Mexico itself. As a way to challenge the commodity paradigm of corn that policy-makers and business

---

5   Bimbo bread is the equivalent to Wonder Bread in the United States: sliced white bread. Its consumption is widespread in urban areas and it is more common in higher socioeconomic brackets. However, its consumption usually supplements *tortillas* rather than replacing them.

leaders in Mexico now use, social advocates have drawn on the key principle on which nation-states legitimize their existence: sovereignty.

Food sovereignty is a principle that was launched at the World Food Summit in 1996 by La Via Campesina, an international coalition of local and national organizations that represent peasants, fishers, and agricultural laborers from across the globe. On its website, the organization defines food sovereignty as:

> the right of peoples to healthy and culturally appropriate food produced through sustainable methods and their right to define their own food and agriculture systems. ... Food sovereignty prioritizes local food production and consumption. It gives a country the right to protect its local producers from cheap imports and to control production. It ensures that the rights to use and manage lands, territories, water, seeds, livestock and biodiversity are in the hands of those who produce food and not of the corporate sector. (La Via Campesina 2011)

In short, it asserts the right for citizens to determine what food they should grow and how they should grow it without foreign influence. It recognizes the connections that exist between social conditions, cultural perspectives, the maintenance of heritage products, and food production.

Using food sovereignty as their rallying principle, many of the civic organizations concerned with the future of corn in Mexico, as well as those advocating for farmers, the environment, and consumers, formed a movement called *El campo no aguanta más* (The Countryside Can Take No More) in 2002. Members participated in blockades on highways and on the border, in land occupations, and in political protests. They also launched broad propaganda campaigns to mobilize the public to support policy that promotes food sovereignty.

Activist efforts have had limited success on the policy front. With pro-business presidents from the National Action Party (PAN) in power from 2000 to 2012, it was highly unlikely that major changes would come about. In fact, the government approved the requests by Monsanto, Dow, and DuPont to plant small experimental plots with transgenic corn. These companies have applied for permits to plant large extensions of these crops with the goal of having the first commercial plot planted by the end of 2012 (Reuters 2011). It is clear that the government does not intend to slow or stop the spread of GMO corn into Mexico. However, 2012 was an election year that returned power to the old ruling party, and the new administration could change the direction of future policy.

The small farmers' activism in defense of Mexican corn, the public campaign against GMOs, and the calls for food sovereignty have had an effect on public consciousness. Corn has become a revitalized national symbol of cultural heritage and its consumption represents an act of patriotism that transcends socioeconomic categories. Nowhere is this more evident than in the proliferation of gourmet restaurants that celebrate "traditional" Mexican cuisines that emphasize indigenous ingredients. Wealthy and middle class Mexicans now find themselves eating traditional dishes featuring *escamoles, chapulines, chinicuiles, chicatanas, huitlacotche, nopalitos,* and iguana, all with fresh handmade *tortillas* and washed down with *pulque.*[6] Until recently, most of these foods were unknown to most middle- and upper-class urban Mexicans, or at the very least they were not considered appetizing. These foods had typically been seen as something that only rural, indigenous people would eat. Moreover, Mexicans are rediscovering the over 600 ways that corn can be consumed (González and Chauvet, 2008). The new popularity of these foods reflects the embracing of the rhetoric of food as culture and food as heritage in Mexico. And no food is more important to Mexican culture than corn.

Conclusion

As corn began to travel around the world, its economic, social, and cultural role began to vary. It became the central metaphor for the creation myths in Mesoamerican societies, creating an essential link between nature and culture. In the Columbian exchange it became a commodity for Europeans and a cultural icon of resistance in the Americas. In the U.S., it became symbolic of the belief in progress as the taming and conquest of nature through science and capitalism. The U.S. tried to impose this vision of corn on the rest of the world by presenting corn as a commodity that Americans could produce more efficiently. Thus, other countries should emulate the United States in its approach to the production of corn, and ultimately cede production to them. This vision was at first embraced in Mexico as it tried to feed its growing population. However, seeing corn as a commodity controlled through science came into conflict with the symbol of corn as a central part of Mexico's heritage and culture. Corn represented one of the last connections to nature in a quickly modernizing world. The advent of GMO corn represented for one side a triumph of science and capitalism over nature and for the other the potential destruction of culture through the corruption of nature.

Corn lives through the meaning that people embed in it. Through its history and journey across the globe, corn has acquired multiple lives, and these lives will continue to clash and come into conflict. Science will continue to nurture and to destroy, promoting some potential futures and foreclosing others. And the Mexican people will continue to care for, nurture, and revitalize corn as the material and symbolic substance of their own lives.

---

6   What Mexicans are eating are ant larvae, grasshoppers, red worms from the maguey plant, flying ants, corn fungus, and cactus with the fermented sap of the maguey plant to drink. It must be pointed out, however, that many of these dishes and ingredients are "legitimized" through the culinary magic of trained chefs and are often interpretations rather than the reproduction of traditional dishes (see Ayora Díaz 2012 for an examination of these trends in restaurants in Yucatán).

References

Ayora-Díaz, S.I. 2012. *Foodscapes, Foodfields, and Identities in Yucatán*. New York: Berghan Books.

Bourdieu, P. 1987. *Distinction: A Social Critique of the Judgment of Taste*. Translated by Richard Nice. Cambridge, MA: Harvard University Press.

Baker, A. 2006. "Ethanol reshapes the corn market." *Amber Waves*. 5(2). Available at: http://www.ers.usda.gov/amberwaves.

Bartra, R. 2002. *Blood, Ink, and Culture: Miseries and Splendors of the Post-Mexican Condition*. Durham: Duke University Press.

Carrasco, M.D., K.M. Hull and R. Wald. 2004. *Unaahil B'aak: The Temples of Palenque*. http://learningobjects.wesleyan.edu/palenque/.

Castellanos, E. and S.M. Bergstresser. 2006. "Food Fights at the EU Table: The Gastronomic Assertion of Italian Distinctiveness." In *Food, Drink and Identity in Europe*, edited by Thomas M. Wilson, 179–202. Amsterdam: Rodopi.

Commission for Environmental Cooperation. 2004. *Maize and Biodiversity: The Effects of Transgenic Maize in Mexico*. Montreal: The Secretariat of the Commission for Environmental Cooperation.

Crane, H. 2006. "God is in the Gluten: Communion as a Test of Faith for Catholics with Celiac Disease." Paper presented at the Society for the Anthropology of Religion and the American Ethnological Society Meetings. Providence, Rhode Island. April 24–26.

Doebley, J. 2004. "The Genetics of Maize Evolution." *Annual Review of Genetics*. 38: 37–59.

Douglas, M. 1966. *Purity and Danger: An Analysis of Concepts of Pollution and Taboo*. New York: Praeger.

Elias, N. 1978. *The Civilizing Process*. New York: Urizen Books.

Fitting, E. 2001. *The Struggle for Maize: Campesinos, Workers, and Trangenic Corn in the Mexican Countryside*. Durham: Duke University Press.

Fussell, B. 1992. *The Story of Corn*. Albuquerque: University of New Mexico Press.

González, R.L. and M. Chauvet. 2008. "Controversias y partcipación social en bioseguridad en México: El case del maíz trangénico." In *Desde los colores del maíz; una agenda para el campo mexicano*, edited by J. Luis Seefoó Luján, 199–232. Zamora, Michoacán: El Colegio de Michoacán.

Jacobsen, E. 2004. "The Rhetoric of Food: Food as Nature, Commodity and Culture." In *The Politics of Food*, edited by Marianne Elisabeth Lien and Brigitte Nerlich, 59–78. New York: Berg.

La Via Campesina. 2011. "What is La Via Campesina? The International Peasants' Voice." [Online]. Available at: http://viacampesina.org/en/index.php?option=com_content&view=category&layout=blog&id=27&Itemid=44 (Accessed December 4, 2011).

Lindholm, C. 2008. *Culture and Authenticity*. Malden, MA: Blackwell Publishing.

McAfee, K. 2008. "Beyond Techno-Science: Transgenic Maize in the Fight over Mexico's Future." *Geoforum* 39: 148–60.

Mintz, S. 1986. *Sweetness and Power: The Place of Sugar in Modern History*. New York: Penguin Books.

Nadal, A. and T.A. Wise. 2004. "The Environmental Costs of Agricultural Trade Liberalization: Mexico–US Maize Trade Under NAFTA." Working Group on Development and Environment in the Americas Discussion Paper Number 4, Tufts University, Medford, MA. Available at: http://ase.tufts.edu/gdae.

Nash, J. 2001. *Mayan Visions: The Quest for Autonomy in an Age of Globalization*. New York: Routledge.

Pilcher, J.M. 2002. "Industrial *Tortillas* and Folkloric Pepsi: The Nutritional Consequences of Hybrid Cuisines in Mexico." In *Food Nations: Selling Taste in Consumer Societies*, edited by Warren Belasco and Philip Scranton, 222–39. New York: Routledge.

____. 2006. *Food in World History*. New York: Routledge.

____. 2008. "Taco Bell, Maseca, and Slow Food: A Postmodern Apocalypse for Mexico's Peasant Cuisine?" In *Food and Culture: A Reader* [Second Edition], edited by Carole Counihan and Penny Van Esterik, 400–410. New York: Routledge.

Pollan, M. 2006. *The Omnivore's Dilemma: A Natural History of Four Meals*. New York: The Penguin Press.

Reader, J. 2011. *Potato: A History of the Propitious Esculent*. New Haven: Yale University Press.

Reuters. 2011. "Mexico Set to Expand GMO planting Group: More than 10 permits sought again for pilot projects." September 19.

Richard, A. 2008. "Withered Milpas: Governmental Disaster and the Mexican Countryside." *Journal of Latin American and Caribbean Anthropology* 13(2): 387–413.

Suárez Carrera, V. 2010. "El campo no aguanta más. Salvemos al campo para salvar a México." *Cuadernos del Movimiento*. Available at: http://www.scribd.com/doc/36140030/El-campo-no-aguanta-mas.

Tedlock, D. 1996. *Popol Vuh: The Definitive Edition of The Mayan Book of The Dawn of Life and The Glories of Gods and Kings*. New York: Touchstone.

Warman, A. 2003. *Corn & Capitalism: How a Botanical Bastard Grew to Global Dominance*. Translated by Nancy L. Westrate. Chapel Hill: University of North Carolina Press.

Weis, T. 2007. *The Global Food Economy: The Battle for the Future of Farming*. New York: Zed Books.

Westcott, P.C. 2007. *Ethanol Expansion in the United States: How Will the Agricultural Sector Adjust?* Darby, PA: Diane Publishing Books.

Chapter 14

Cultural Heritage in Food Activism: Local and Global Tensions

Carole Counihan

This chapter explores the role of cultural heritage in food activism in Italy through an ethnographic case study from the city of Cagliari on the island of Sardinia. It examines whether recourse to culinary cultural heritage can advance the overriding goal of food activism: the promotion of food democracy. I discuss the constantly evolving definition of cultural heritage in food activism, its potential to promote economic justice and cultural survival, and cultural heritage as a force of inclusion and exclusion.

Cagliari is the capital of the island region of Sardinia and the place where I did my first anthropological fieldwork in 1976. I have continued doing ethnographic fieldwork in Italy since then with projects in 1978–79 in Bosa (Sardinia), in 1982–84 and 2003 in Florence, and all over Italy in 2009 studying Slow Food. I was a visiting professor at the University of Cagliari from April–June, 2011, and I studied Cagliari's alternative food scene by conducting approximately 50 interviews and observations, including 17 recorded interviews, with various participants. With 156,000 residents in the city and 563,000 in the surrounding metropolitan area, Cagliari boasts a third of the island's total population of 1,675,000 and is its major port.[1] It is also the region's key cultural, political, commercial, transportation, and tourist hub. It is a good place to study the contemporary Italian alternative food movement because of Sardinia's strong agro-pastoral roots, its rapidly changing foodways, its high density of supermarkets and discount food stores,[2] and its burgeoning alternative food scene. I was able to study two recently started farmers markets, several organic producers and distributors, the active Slow Food chapter, a vegetarian restaurant, farm-to-school alliances, and the GAS (Gruppo di Acquisto Solidale) Cagliari Circolo Aperto, the Open Circle Solidarity Buying Group.

The GAS movement started in 1994 in Fidenza (Parma) and has expanded throughout Italy (Grasseni 2012, Grasseni 2013, http://www.retegas.org/). It consists of a loosely organized network of over 600 independent local groups

---

1   For Italy population figures, see http://demo.istat.it/pop2011/index.html.

2   See Burresi 2002 on distribution of supermarkets in Italy through 2000. For recent statistics, see the supermarket industry website: http://www.infocommercio.it/pagine/banche-dati/index_supermercati.php.

seeking greater justice and quality in the food system. Consumers join together to buy directly from local producers. Members pay a small annual fee and can make a weekly order of all the products the GAS offers, which producers bring to a central location for GAS members to pick up. This chapter uses interviews and observations with the GAS Cagliari president and founding member Lucio Brughitta and other food activists to explore the connections between their activist efforts and recourse to concepts of cultural heritage.

Food activism refers to efforts to promote social and economic justice through the food system. In Italian, I did not use the literal translation of food activism, "*attivismo alimentare*" to describe my project, since the concept in Italian has extremist connotations and did not accurately describe my focus. Instead, I told people I was studying "*l'alimentazione alternativa*"—"alternative foodways"—those challenging the agro-industrial model and aiming for food democracy—universal access to high quality, affordable, sustainable, and culturally appropriate food (Lang 1999). Food democracy also means including "all the voices of the food system" (Hassanein 2003: 84). In cultural heritage discourse, we can ask whose voices prevail and whose are missing?

Cultural heritage expressed as food is a complex moving target, manipulated by diverse people for diverse ends. It is both tangible and intangible, and falls under the UNESCO Convention on Intangible Cultural Heritage (UNESCO 2003), which states: "The 'intangible cultural heritage' means the practices, representations, expressions, knowledge, skills—as well as the instruments, objects, artefacts and cultural spaces" of social groups passed down through generations that contribute to identity, are grounded in place, and are 'a mainspring of cultural diversity and a guarantee of sustainable development'" (Art. 2, point 2, p. 3). Cultural heritage in food includes the material: landrace plants, traditional dishes, tools, landscapes, and so on; as well as the immaterial: cuisine, intellectual and corporeal knowledge, traditions and techniques, ideology, sensory awareness, philosophies of food and health, and so on. Following Sims' (2009: 324) work on culinary authenticity in tourism, in food activism we can ask how "people make claims" for cultural heritage and "the interests that those claims serve."

In my research on food activism in Cagliari, people rarely spoke of *patrimonio culturale* (cultural heritage), perhaps because I never introduced these terms, or perhaps because other concepts were more meaningful. In Italy and Sardinia, cultural heritage in food is intimately embedded in history, identity, community, and place. Cagliari food activists talked about local foods, Sardinian foods, regional or village foods, and foods of the *territorio*—by which they meant their home, their locality, their meaningful natural and cultural place. Their understandings of products of the *territorio* were, however, anything but uniform or static. For example, when I asked the Terra di Mezzo vegetarian restaurant co-owner and chef Lara Ferraris, "Do Sardinian things have a place in your cuisine?" she replied:

> All the vegetables are absolutely local. The producers are from nearby and we always try for zero kilometers. As for the really traditional Sardinian

recipes—the Sardinian cuisine has a lot of meat, a lot of fish, so basically, not really. Maybe, I don't know, we make *maloreddus* [Sardinian gnocchi made out of semolina flour], ... the typical pasta, there's no problem, you can dress it any way. For example, there are the classic *maloreddus alla campidanese*, the ones we eat in Cagliari with sausage. Obviously here we don't make them with sausage, however, we can make them with a seitan [wheat gluten] sauce, so we can always adapt the recipe to our cuisine .... (Ferraris interview 10 May 2012)

Ferraris defined Sardinian cuisine, an important form of cultural heritage, as meaty but malleable, and I submit that the case of meat reflects the broad flexibility of cultural heritage definitions in food, which evolve over time and reflect changing socio-economic conditions. While meat has certainly had a place in Sardinian cuisine, it has not been the dominant food of most people for most of Sardinian history (Angioni 1974, 1976, Cirese et al. 1977, Guigoni 2009, Le Lannou 1941). On the contrary, as recently as 1978–1979 when I did fieldwork in the town of Bosa on changing food habits, I found a complex vegetarian cuisine rooted in poverty containing myriad dishes based on chickpeas, beans, favas, peas, lentils, eggplant, fennel, artichokes, tomatoes, potatoes, mushrooms, spinach, Swiss chard, onions, garlic, parsley, carrots, celery, cabbage, cauliflower, and many kinds of wild greens (Counihan 1981). The diet of many Bosans had been largely vegetarian by necessity, like that of many urban and rural poor in Italy for centuries, because animal products were expensive (Capatti, De Bernardi and Varni 1998, Capatti and Montanari 1999, Conti 2008, Counihan 2004, Helstosky 2003, Sorcinelli 1998, Vercelloni 1998, 2001, Zamagni 1998). Sardinia has great diversity of climate, land, and cuisine, from the central mountainous sheep-rearing regions to the fishing community of S. Elia near Cagliari to the vegetable gardens grown almost everywhere in the past, so regional culinary diversity abounds with varying emphasis on vegetables, legumes, fish, and meat complementing the dietary centerpiece—bread (Counihan 1984). Even in the Barbagia region notorious for pastoralism, Alessandra Guigoni (2009) has found a rich horticultural and culinary presence of vegetables, legumes, and fruits.

Like so many other populations, with a rise in their standard of living, Sardinians and all Italians have witnessed a significant increase in meat consumption in the twentieth century (Vercelloni 2001), which may partly explain why some today define meat as Sardinia's culinary heritage and why Slow Food and other activists have worked to promote Sardinian breeds of cattle. The Slow Food Cagliari chapter supported butcher Walter Vivarelli who sold meat only from the *Consorzio del Bue Rosso della Razza Sardo-Modicana* (Consortium of the Red Steer of the Sardo-Modicana Breed), which is a Slow Food presidium product.[3] Slow Food presidia and "*comunità del cibo*," (food communities) seek to acknowledge,

---

3   On the Slow Food presidium of the Sardinian red steer of the Sardo-Modicana breed, see http://www.presidislowfood.it/ita/dettaglio.lasso?cod=36, accessed 14 December 2011.

protect, and add value to traditional high quality regional foods—forms of cultural heritage (Sassaelli and Davolio 2010, Siniscalchi 2011).

Slow Food Cagliari obtained presidium Sardo-Modicana beef from butcher Vivarellli for its twentieth anniversary dinner held beneath the stars in the Monte Claro Park in Cagliari on June 18, 2011. After a first course of *ciccioneddos alla campidanese* (small gnocchi-shaped semolina pasta served with meat sauce), there were three beef dishes all from the *bue rosso*: *bombas* (meatballs), *spezzatino* (beef stew), and *spezzatino nero* (black beef stew) made with Guinness stout. The idea was, as Vivarelli had said at a symposium earlier that day, to reverse the situation of today where "Sardinians themselves no longer know the products of their own land—*i prodotti della propria terra.*" Traditionally, meat has marked a feast and the Slow Food event continued that tradition in a new form while supporting Sardinian cattle producers. Culinary heritage is constantly evolving through education and renewed practices, but it maintains a consistent focus on local products developed in the territory over time associated with regional communities and identities.

Another example of recourse to cultural heritage was Slow Food Cagliari's establishment in June 2010 of a *"comunità del cibo"* for the caper to support revival of its production, which for 100 years had been a major agricultural industry in Selargius, five kilometers from Cagliari, and had almost become extinct due to competition from capers cheaply produced in North Africa (Guigoni 2010). Widespread in the circum-Mediterranean, capers are the buds of the caper bush which are usually pickled in brine or vinegar. On June 12, 2011, Slow Food Cagliari held the first anniversary celebration of the caper food community. The event included several speeches about the history and current status of local caper production; a visit to a nearby field where caper plants were growing; and a guided caper tasting among guests seated at convivial tables with plenty of wine to wash down the salty, vinegary capers. Promotion of the Selargius capers emphasized their long history in the area, their superior taste, their economic importance, and their embeddedness in local land, cuisine, and culture. The revival of caper production contributed to cultural and plant survival. Its contribution to food democracy may be more difficult to affirm because it aimed to provide a just income for producers, but that meant selling the capers at 23 euros per kilogram and twice that for smaller quantities—considerably more expensive than cheap imports. This case illustrates an ongoing tension between economic justice and recourse to culinary cultural heritage, which may bring a living wage for producers but at prices too high for many consumers.

Clare Hinrichs (2003: 33) poses the important question of whether a focus on local foodways is "liberatory" or "reactionary." DuPuis and Goodman (2005: 360) offer one answer when they say that localism "can provide the ideological foundations for reactionary politics and nativist sentiment." In 2009, the Tuscan town of Lucca enacted a ban on new ethnic food shops in the historical center, which stirred this very debate. Some argued that was an example of nativism and exclusionary politics; others argued it was a legitimate protection of local cultural heritage (Donadio 2009). Similarly, in 2010, the Tuscan beach town of Forte di Marmi said "no to American fast food, no to kebabs and sushi bars in the historic center or beachfront." The *"genius loci"* decree was not xenophobia, said the mayor, but "a necessary regulation to defend our cultural identity" (Montanari 2011a, my translation). Hostility towards immigrants abounds in Italy as social scientists have documented (Daly 1999, Randall 2009, Riccio 1999), and as I was writing this chapter in December 2011, an Italian with ties to the far right murdered two Senegalese immigrants and wounded several others at an open air market in Florence; the local imam called this act "fruit of ten years of politics of hate against immigrants" (Montanari 2011b). It is interesting to note that these three incidents all occurred in tourist hotspots which depend on marketing their cultural identity, authenticity, and heritage to outsiders, perhaps explaining their zealous culinary protectionism.

An emphasis on local foods can isolate, render suspect, and delegitimize the foods of Italy's growing numbers of immigrants—approximately four million, seven percent of the population (Randall 2009: 16). In spite of the huge influence on Italian cuisine of such immigrant plants as tomatoes, potatoes, corn, and beans (Guigoni 2009), Cinotto points out that "the Italian culinary model seems to resist almost completely the influences of immigrant cuisines" (2009: 670, my translation). He attributes this resistance to both Italian racism and to the economic benefits of emphasizing local foods, benefits that Brasili and Fanfani's (2010) statistical analysis of agrifood districts supports.

Extreme localism can also erase immigrants' important contribution to the farming, harvesting and processing of local foods—from the Pakistani and Moroccan butchers preparing prosciutto in Parma, to Sikhs raising and milking cattle in the Val Padana, to Romanians and Albanians herding sheep in the Abruzzo and Sardinia (Cinotto 2009, Contu 2011, Randall 2009, Sias 2011). An important goal of the alternative food movement is to connect producers and consumers but immigrants rarely appear in that connection because they are more likely to be laborers working in the fields than farm-owners who meet consumers at farmers markets or farm visits. Lucio Brughitta of the GAS Cagliari said their group regularly organizes trips to the small producers that supply the GAS:

> The visit to a producer involves the visit, learning about the farm, the lands of the farm, the ways of working in that farm. Going to visit a producer means speaking with the producer, asking him about his life, why he has chosen to operate in a certain way rather than another, to dedicate himself to quality production rather than quantity production, what are the problems he has with his work, the problems with his family. This helps us a lot because we establish a relationship of trust between us the consumers and him the producer. (Brughitta interview 20 June 2011)

Like the GAS members, many food activists aim for close ties with producers but very often focus on independent farmers and/or artisans, and the agricultural proletariat gains less attention or is ignored (Wald 2011). For example, I joined Slow Food Cagliari on a visit to the Santa Margherita di Pula Tomato Cooperative Sapore di Sole where we had an organized tour of the greenhouse and packing plant.[4] We spoke with the public relations director and the scientists, but not with any of the 100 female workers at the packing plant. At the Mercato Campagna Amica farmers' market organized by the Coldiretti, Italy's largest farmers' union, I talked several times with Mr. and Mrs. Ortu, independent vegetable farmers. They spoke of how wholesalers squeeze farmers, how prices have dropped while expenses and regulations have risen, how hard it is to make it in farming, how difficult it is to compete against cheaper produce from northern Africa, but nonetheless how much they want to continue producing high quality Sardinian crops.

At the market, I also spoke with two Coldiretti farmers' union employees about farmworkers, among other topics. They told me that Sardinia has Romanian, Indian, and Senegalese farmworkers who almost always have visas and work permits because the fines for hiring illegal immigrants are steep. However, recent race riots in Calabria, Italy between Italians and undocumented migrant farmworkers from the Maghreb and sub-Saharan Africa highlighted the use of illegal immigrant labor, the exploitative conditions and low pay in agriculture, and the extreme racial hostility of some Italians towards immigrants (Immigration 2010). Anthropologists at the University of Cagliari have found immigrants marginalized in Sardinia whether legal or not (Bachis and Pusceddu 2011, Contu 2011, Sias 2011). Issues of labor and class, race-ethnicity, and gender need greater sustained attention in food activism so that all the voices of the food system are heard.

Hinrichs (2003: 36) offers a useful perspective on thinking about cultural heritage and food democracy when she uncovers two tendencies in food activism: "defensive localization and more diversity-receptive localization." In Cagliari, there was evidence of both. For example, the very name of the Cagliari GAS, "Circolo Aperto" or "Open Circle," proclaimed an inclusive ideology reflected in diverse practices. One of their main suppliers of fruits and vegetables was a female German immigrant biodynamic farmer. Furthermore, Brughitta explicitly rejected the extreme localism known in Italy as *campanilismo* or "bell-tower-ism" and made room for products of excellence from elsewhere when I asked him about the importance of *territorio*:

> It is fundamental. In our GAS, we privilege local producers, hence producers from the *territorio*, the typical products of the territory, the most healthy, what we can still today succeed with some sacrifice in procuring but which must survive. But clearly we don't have problems of *campanilismo* with food, some things we seek outside of the island, for example, products of absolute

---

4   The Coop's website says they also organize tours for schools, with the aim of explaining greenhouse tomato production to students, but there was no evident effort on the website to entice tourists. See http://www.smargherita.it/, accessed 2 July 2012.

excellence like *parmigiano reggiano* have a production area located in northern Italy, and we don't have that type of production, so with complete tranquility we buy *parmigiano reggiano* ... We privilege local products because we have the goal of privileging local producers, because the farm family has to have a just income. Otherwise, we will cease to exist as Sardinians. (Brughitta interview 20 June 2011)

Brughitta linked localism to quality food, economic justice, and cultural survival of Sardinians, but did not preclude purchasing foods from other locales or made by non-Sardinian producers.

Another food activist, Teresa Piras, a founder and leader of the *Centro di Sperimentazione Autosviluppo*[5] (the Center for Self-Development Experimentation) described several initiatives to promote local development by creating a "tie between consumers and products of the *territorio*." Her group valued, used, and constantly redefined cultural heritage in food. For example, they fostered an event to graft heirloom pears onto wild pear trees. The group organized a workshop to build an *orto sinergico*, a raised-bed no-till garden to grow vegetables fundamental to Sardinian cuisine, and another on how to bake traditional bread using a sourdough starter (*pasta madre*) passed down from the baker's grandmother to share with all the participants. These workshops highlighted education about local Sardinian cultural heritage in revitalizing knowledge about fruit, vegetables, and grains that had major importance to the diet, agriculture, material arts, and survival. Their initiatives could potentially foster food democracy by making available healthy, sustainable, and culturally meaningful foods at accessible prices.

Like Brughitta, Teresa Piras rejected what Hinrichs would call defensive localism as is evident in the following excerpt from my fieldnotes (30 May 2011):

> Teresa also mentioned how they have 'a kind of GAS' that is coordinated with their weekly meetings because they wanted to bring the 'concept of fair trade here to Sardinia.' So they have the meeting, and they 'shop together' by ordering ahead from a local farmwoman and having her deliver the produce to their weekly meeting. And then they have fair trade products that are not produced in the area: coffee, tea, sugar, and especially chocolate—'because we like it and it's good!' Teresa said. They buy fair trade products from UCIRI (Union of Indigenous Communities of the Isthmus Region)[6] and CTM—Cooperazione Terzo Mondo[7]—to keep on hand with the aim of providing 'support for small producers' for products that Sardinia does not have. She said that they have these

---

5 For more on the *Centro di Sperimentazione Autosviluppo*, see the group's website: www.domusamigas.it.

6 See UCIRI's website: http://www.uciri.org/english/english.htm, accessed 7 December 2011.

7 See Altromercato, the CTM website: http://www.altromercato.it, accessed 7 December 2011.

products because their work 'is not a form of closure, but of solidarity with all small producers ... we want to promote the sharing of wealth.'

Teresa Piras valued the cultural heritage foodways of the small Sardinian farmers from whom her group bought local produce, and of the small farmers in Oaxaca, Mexico from whom they bought their coffee, linking them in a commitment to economic justice.

Not only is cultural heritage a means of promoting small farmers and their products locally and globally, it is leverage for education, good healthy food, and environmental sustainability. For example, organic food producer and distributor Matteo Floris told a story about supplying a local nursery school run by the Province of Cagliari with organic produce. He said that the administration there ordered from him a thousand kilograms of bananas and 50 kilograms of oranges. He explained to the nursery school dietician and director that he could procure organic bananas, but they came from hundreds of miles away, whereas he himself grew the oranges on Sardinian soil. Thus he tried to encourage them to order more oranges and fewer bananas to support the local agricultural economy, Sardinian traditional diet, and sustainable food. But how might such initiatives affect the food preferences of immigrant children? Depending on where they come from, their foods might not be available locally. With immigrants totaling 7 per cent of Italy's population and growing, what will happen to their foodways? While food activists support small farmers and diverse foodways globally, will they support them on Italian soil? Will a discourse of cultural heritage make room for immigrant foodways? When immigrant hands grow them, do foods retain their cultural heritage?

There is some evidence that although immigrant foodways are resisted by mainstream Italians, they are finding a niche among other immigrants and young people (Donadio 2009). Several mini-ethnography projects by Masters students in my Food Anthropology classes at the University of Gastronomic Sciences in Colorno/Parma and Pollenzo/Bra on mainland Italy over the last seven years have shown over and over again that *extracomunitari*—immigrants from outside the EU, including African, middle eastern, Chinese, and North American—often shopped and ate at businesses run and frequented by immigrants of their own and other groups, including kebab shops, Halal butchers, and "ethnic" grocery stores—locales which were inexpensive, familiar, comfortable, and open convenient hours. But bans like that of Lucca, prohibiting ethnic food stalls from the city center, may effectively keep immigrants out as well. Research on Senegalese in northern Italy shows that for immigrants contested foodways can symbolize the struggle to find a place in Italian society (Cinotto 2009, Gasparetti 2012, Riccio 1999). Italians' strong attachment to their regional cuisines combined with the newness of immigration mean that the future of immigrants and their foodways in Italy is a provocative and unresolved issue.

Cultural heritage discourse in food activism is mobilizing support for local products, economies, and traditional knowledge about foodways, however it is not so easily inclusive of non-local people, cuisines, and cultures, nor of the laborers who produce the local foods. More research is needed on all stages of production of cultural heritage foods to determine if there are fair working conditions and just remuneration. Food activism imagines a plurality of localisms—both in Italy and around the globe, but needs to engage more directly with the increasing cultural diversity of foodways in Italy at the same time as promoting culinary cultural heritage and food democracy.

References

Angioni, G. 1974. *Rapporti di produzione e cultura subalterna: Contadini in Sardegna*. Cagliari: EDES.

Angioni, G. 1976. *Sa Laurera: Il lavoro contadino in Sardegna*. Cagliari: EDES.

Bachis, F. and A. Pusceddu. 2010. Mobilità, confini, religioni: per un approccio comparativo in area mediterranea. Talk at the University of Cagliari, May 16, 2011.

Brasili, C. and R. Fanfani. 2010. The Agrifood Districts in the New Millennium. *Proceedings in Food Systems Dynamics. System Dynamics and Innovation in Food Networks*, edited by M. Fritz, U. Rickert, G. Schiefer, pp. 10–30.

Capatti, A. and M. Montanari. 1999. *La cucina italiana: storia di una cultura*. Roma-Bari: Laterza.

Capatti, A., A. De Bernardi and A. Varni (eds) 1998. *Storia d'Italia: l'alimentazione*. (Annali 13). Torino: Einaudi.

Cinotto, S. 2009. "La cucina diasporica: il cibo come segno di identità culturale," in *Storia d'Italia. Annali 24. Migrazioni*. Paola Corti and Matteo Sanfilippo (eds). Turin: Einaudi, pp. 653–72.

Cirese, A.M., E. Delitala, C. Rampallo and G. Angioni. 1977. *Pani tradizionali, arte effimera in Sardegna*. Cagliari: EDES.

Conti, P.C. 2008. *La leggenda del buon cibo italiano e altri miti alimentari contemporanei*. Lucca: Fazi.

Contu, S. 2011. Drum bun! Sistema migratorio rumeno: vecchi e nuovi contesti di mobilità. Talk at the University of Cagliari, May 19, 2011.

Counihan, C. 1981. *Food Culture and Political Economy: Changing Lifestyles in the Sardinian town of Bosa*. Doctoral Dissertation, Anthropology, University of Massachusetts.

Counihan, C. 1984. Bread as World: Food Habits and Social Relations in Modernizing Sardinia. *Anthropological Quarterly*, 57, 2: 47–59.

Counihan, C. 1988. Female Identity, Food, and Power in Contemporary Florence. *Anthropological Quarterly* 61, 2: 51–62.

Counihan, C. 1999. *The Anthropology of Food and Body: Gender, Meaning and Power*. New York: Routledge.

Counihan, C. 2004. *Around the Tuscan Table: Food, Family and Gender in Twentieth Century Florence*. New York: Routledge.

Daly, F. 1999. Tunisian Migrants and their experiences of Racism in Italy. *Modern Italy* 4, 2: 173–89.

Donadio, R. 2009. Lucca Journal: A Walled City in Tuscany Clings to Its Ancient Menu. *New York Times*, March 12, 2009 http://www.nytimes.com/2009/03/13/world/europe/13lucca.html (consulted 15 December 2011).

DuPuis, E.M. and D. Goodman. 2005. Should We Go "Home" To Eat? Toward a Reflexive Politics of Localism. *Journal of Rural Studies* 21: 359–71.

Gasparetti, F. 2012. Eating a *tie bou jenn* in Turin: Negotiating differences and building community among Senegalese migrants in Italy. Forthcoming in *Food and Foodways*, 20, 3–4.

Grasseni, C. 2012. Reinventing food: the ethics of developing local food. In *Ethical Consumption: Social Value and Economic Practice.* James G. Carrier and Peter G. Luetchford (eds). Oxford: Berghahn Publishers.

Grasseni, C. Forthcoming. Dallo home-banking allo home-baking. Sobrietà, sostenibilità e solidarietà nelle nuove culture del pane. In *Cibo e Sacro. Culture a confronto.* Luigi M. Lombardi Satriani e Roberto Cipriani (eds). Roma: Armando Editore.

Guigoni, A. 2009. *Alla scoperta dell'America in Sardegna. Vegetali americani nell'alimentazione sarda.* Cagliari: AM&D.

Guigoni, A. 2010. Il cappero di Selargius. Aspetti storici e culturali di una pianta ultracentenaria. *Anthropos e Iatria*, 14, 4: 8–13.

Hassanein, N. 2003. Practicing Food Democracy: A Pragmatic Politics of Transformation. *Journal of Rural Studies* 19: 77–86.

Helstosky, C.F. 2004. *Garlic and Oil: Politics of Food in Italy*. New York, Oxford: Berg Publishers.

Hinrichs, C.C. 2003. The practice and politics of food system localization. *Journal of Rural Studies* 19: 33–45.

Immigration in Italy, Southern misery: An ugly race riot reflects social tensions and economic problems in the south. *The Economist*, 14 January 2010.

Lang, T. 1999. Food Policy for the 21st Century: Can It Be Both Radical and Reasonable? In *For Hunger-Proof Cities: Sustainable Urban Food Systems.* M. Koc, R. MacRae, L.J.A. Mougeot and J. Welsh (eds). Ottawa: International Development Research Centre, pp. 216–24.

LeLannou, M. 1941. *Pâtres e Paysans de la Sardaigne.* Tours: Arrault e C.

Montanari, L. 2011a. Forte dei Marmi dice stop a kebab e fast food. *La Repubblica*, 9 October 2011. http://firenze.repubblica.it/cronaca/2011/10/09/news/forte_dei_marmi_dice_stop_a_kebab_e_fast_food-22957601/index.html?ref=search accessed 8 July 2012.

Montanari, L. 2011b. Firenze, spari contro i senegalesi, estremista fa due morti e si uccide. *La Repubblica*, 13 December, 2011. http://www.repubblica.it/cronaca/2011/12/13/news/firenze_gianluca_casseri_killer_senegalesi_suicida-26554634/ accessed 14 December 11.

Randall, F. 2009. Italy's New Racism. *The Nation,* 2 February 2009, 16–17.

Riccio, B. 1999. Senegalese Street-sellers, racism, and the discourse on "irregular trade" in Rimini. *Modern Italy* 4, 2: 225–39.

Sassatelli, R. and F. Davolio. 2010. Consumption, Pleasure, and Politics: Slow Food and the politico-aesthetic problematization of food. *Journal of Consumer Culture* 10, 2: 1–31.

Sias, C.G. 2011. Tra Sardegna e Albania. Riflessioni per una etnografia multisituata. Talk at the University of Cagliari, May 30, 2011.

Sims, R. 2009. Food, Place and Authenticity: Local Food and the Sustainable Tourism Experience. *Journal of Sustainable Tourism*, 17, 3, May 2009, 321–36.

Sorcinelli, P. 1998. Per una storia sociale dell'alimentazione. Dalla polenta ai crackers. In *L'alimentazione*. Alberto Capatti, Alberto De Bernardi, e Angelo Varni (eds). Torino: Einaudi, pp. 453–93.

UNESCO 2003. UNESCO Convention for the Safeguarding of Intangible Cultural Heritage. Available at: http://www.unesco.org/culture/ich/index.php?lg=en&pg=00006, accessed 30 November 2011.

Vercelloni, L. 1998. La modernità alimentare. In *Storia d'Italia, Annali 13. L'alimentazione*. Alberto Capatti, Alberto De Bernardi and Angelo Varni (eds). Torino: Einaudi, pp. 951–1005.

Vercelloni, L. 2001. Le abitudini alimentari in Italia dagli anni ottanta agli anni duemila. *Sociologia del Lavoro*, 83, pp. 141–9.

Wald, S.D. 2011. Visible Farmers/Invisible Workers: Locating Immigrant Labor in Food Studies. *Food, Culture and Society* 14, 4: 567–86.

Zamagni, V. 1998. L'evoluzione dei consumi fra tradizione e innovazione. In *Storia d'Italia, Annali 13. L'alimentazione*. Alberto Capatti, Alberto De Bernardi and Angelo Varni (eds). Torino: Einaudi, pp. 171–204.

Index